London's Navy

The City Meets its Navy

On Lord Mayor's Day 1976 the first Naval Reserve Officer ever to be elected Lord Mayor of London, Commander Sir Robin Gillett, Bt., GBE, RD, R.N.R. greets the Commander of London Division watched from the coach by the Sergeant-at-Arms, Lieutenant Colonel St John Brooke Johnson, MBE, whose father joined London Division in 1914 as an Ordinary Seaman and saw service in the Royal Naval Division.

The rum barrel in the foreground was presented to London Division in 1970 by the Worshipful Company of Coopers. (R.N.)

London's Navy

A story of the Royal Naval Volunteer Reserve

Gordon Taylor

Foreword by H.R.H. The Prince of Wales

Quiller Press

Published in 1983 by Quiller Press Ltd
50 Albemarle Street, London W1X 3HE

ISBN 0 907621 18 X

Designed by Jim Reader
Design and production in association with
Book Production Consultants, Cambridge

Printed and bound by
The Burlington Press (Cambridge) Ltd,
Foxton, Cambridge CB2 6SW

Contents

Foreword

H.R.H. The Prince of Wales

BUCKINGHAM PALACE

The volunteer has always had an important part to play in our history.

This is the story of a relatively small and little known band of such volunteers, men and women, who selflessly devote their spare time to learning the increasingly specialised skills required to support the Royal Navy in time of need. Today they have amongst their special concerns the task of helping to ensure that the approaches to our ports are kept free from the threat of mines.

London's Navy is one of eleven 'City Navies' sited in major ports around the country, each of which contributes a vital piece in the complicated jig-saw of seaward defence, and in addition represents the Royal Navy in areas where the permanent service is rarely seen. Overall there are some thirty Reserve units across the Nation all manned by civilian volunteers.

It has often been said in times of crisis or emergency that it is well for us to be reminded that many of the most exceptional human achievements spring simply from a sense of duty fostered by tradition. This book traces the origin of such a tradition and relates how, despite discouragement, it has flourished and gained full recognition by reason of little more than the deep sense of purpose and devotion to the Royal Navy of the individuals concerned.

As one who has some experience of sailing the seas in a mine-sweeper, sharing the good times and the bad, and knowing well the personal qualities required of everyone on board in order to ensure the safe completion of the task, I commend this story to the reader.

Charles

Preface

Admiral of the Fleet The Lord Lewin of Greenwich

'The Regulars keep the Royal Navy going in peace time; when there is a war, the Reserves take over'.

This popular aphorism, not without its grain of truth, was current in World War II, when I observed at first hand the vital part played by the Reserves. Together with their Hostilities Only colleagues, they outnumbered the regulars by more than ten to one, and the Navy could not have operated without them.

After the war many of these men and women with front line naval experience carried on in the Reserves, and their numbers were refreshed as the years went by with those who had completed their National Service in the Royal Navy. This cadre of officers and men with first hand knowledge of the Royal Navy in war and peace provided an invaluable bridge of common understanding with the full time professional Navy.

Time passed, and by the early 1970's, even those with National Service experience were, like old soldiers, fading away. There was a danger that without this link the Royal Navy and the Royal Naval Reserve would grow apart, each not properly aware and perhaps suspicious of the other's role and attitude. The Admiralty Board realised this and, guided by the recommendations of a study that Rear Admiral Mitchell was commissioned to undertake, Their Lordships took the bold step of abandoning the separate status of the R.N.R. under their own Admiral Commanding Reserves and integrating them more fully into the command organisation of the Royal Navy. I was fortunate to be the Commander-in-Chief Naval Home Command at the time, and on 1 January 1977 I took over command of the Royal Naval Reserve.Shortly before, the last Admiral Commanding Reserves, Rear Admiral Hollins, at a ceremony on the quarterdeck of my flagship, HMS *Victory*, had handed to me for safe keeping the R.N.R. officer's sword traditionally worn by 'their' Admiral.

In my remaining time as Commander-in-Chief, and indeed during my spell as First Sea Lord, I came to know the Reserves well, to appreciate even more the unselfish dedication of these patriotic men and women and the contribution so willingly made by their families. But much more important, I watched with both pleasure and satisfaction the growing bond of mutual respect and understanding between today's Navy and its Reserves.

My forty-three years of service have taught me many things about the problems of Defence, above all the difficulty of foretelling the future, and how it is usually the unexpected that happens. Who could have foreseen, even weeks or days before the event, that after many years of training for war as a member of the NATO alliance, we would be sending a Task Force to the South Atlantic to liberate the captured Falklands? Or that my old NATO Headquarters at Northwood, which could not operate in wars or crisis without the men and women of its Reserve Unit, would need their help for such a campaign?

No one can say with certainty what problems of security we will face in five or ten years time. What can be said with certainty is that the best way to prevent war in these coming years is to be ready and to show our determination to defend ourselves. In this, by the time they devote to their training and their readiness to serve their country, the Reserve play a part out of all proportion in importance to their numbers.

Introduction

This is a story of the 'Wavy Navy'. Although those words strictly relate to the R.N.V.R., they typify for the average Briton today the temporary officers who served in the Royal Navy for the duration of the Second World War and proudly wore on their sleeves the wavy gold braid.

What the average Briton may not know is that those who belonged to the 'Wavy Navy' at that time were of two kinds – firstly, the large number of men who, having been accepted for the Royal Navy for the duration of hostilities, became temporary officers and served only until demobilised, and, secondly, the much smaller number of officers and men who joined the R.N.V.R. in years of peace and were called up at the outbreak of war. Within a formal career structure these had been continuously trained in peacetime in its local Divisions throughout Great Britain and they were ready for the call when it came. It is the latter kind with which this book deals – the volunteers who join the Reserve in time of peace and contract to serve in the Royal Navy, afloat or ashore as required, on mobilisation. The wavy braid having already gone, in 1958 the R.N.V.R. dropped the 'V' when it merged with the Royal Naval Reserve, and in 1952 the Women's Royal Naval Volunteer Reserve (now the Women's Royal Naval Reserve) became concerned in the story.

One of the eleven local R.N.R. Divisions is in London, with its two drill ships the *President* and the *Chrysanthemum*, moored in the Thames alongside the Embankment above Blackfriars Bridge, and, although we are dealing with this Division in particular, all Divisions have much the same history. The year 1983 marks the eightieth anniversary of the Reserve. It is a body of men and women dedicated to the service of the nation come what may, and locally is well described as 'London's Navy'.

St Giles-in-the-Fields Rectory,
15A Gower Street,
London, W.C.1.

Gordon Taylor
1 March 1983.

CHAPTER ONE
The Volunteer Heritage

At Trafalgar about twenty-two per cent of the crew of Nelson's *Victory* were volunteers.[1]

That information may surprise those who believe that naval recruitment at that time relied solely on the activities of the press-gangs, but the truth is that in times of national need or emergency the Navy has usually found it necessary to apply the volunteer principle for the proper manning of the fleet, even after the obvious natural reserve of professional seamen has been taken into consideration.

Sir Richard Grenville

From that 'natural reserve' at the time of the Armada came Sir Francis Drake, a lifelong seaman, together with the seamen of the merchant ships appointed to serve under him; but there were also serving in the Queen's Ships many distinguished volunteers who loved a fight but could never be classed as whole-time seafarers. Typical of these must be reckoned George Clifford, Earl of Cumberland, Lord Thomas Howard, Lord Henry Seymour, Lord Sheffield and Sir Robert Southwell, all of whom except Clifford actually commanded the ships in which they served against the Spaniards. Sir Richard Grenville, the Cornish landowner who commanded the *Revenge* in her immortal fight in 1595, and his cousin, Sir Walter Raleigh, both of whom were not serving at the time of the Armada, are two further examples of such volunteers. As Michael Lewis has said, all these were the 'R.N.V.R. officers' of their day; had he been writing after 1958 he would have had to say 'R.N.R.', following the merger of the two Reserves. We should note that Raleigh when in London during the years 1583–1603 lived at Durham House in the Strand which with its gardens used to cover the whole of what is now known as the Adelphi, and which stood only half a mile away from where H.M.S. *President*, the drill ship of London Division of the Royal Naval Reserve, is moored today in King's Reach in the Thames. As Durham House stood opposite the point at which the Thames turns sharply eastwards, and as it therefore had commanding views in both directions, to Westminster and the City of London, Raleigh, in

Sir Walter Raleigh, 1588

his study which was in a turret of the house, must have gazed often at the stretch of water which for more than a century now has seen the activity of London's naval volunteers. Thus the men and women of today who serve in London Division should hold his memory in due esteem.

A volunteer who was serving in one of the thirty ships fitted-out by the City at the time of the Armada emergency, and who is of special interest for the purpose of this history, was John Watts, the owner and provider, but not the commanding officer, of the *Margaret and John*, in which he experienced some of the fiercest fighting. He was knighted in 1603 and became Lord Mayor of London in 1606.[2] These London ships may have carried a fair number of such volunteers, and it is a great pity that the muster-rolls no longer exist.

While volunteers doubtless saw themselves chiefly as fighting men rather than seamen, there were others among them who were clearly neither. In 1598 a clergyman, John Layfield, the Earl of Cumberland's chaplain, accompanied his master's expedition to Porto Rico in the West Indies, and later wrote the account of it which only survives in Purchas's *Pilgrimes*. In 1601 Layfield became Rector of St Clement Danes in the Strand, which is only a quarter of a mile distant from H.M.S. *President*, and of which the parish boundary with St Bride's, Fleet Street, passes through the ship. He remained there till his death sixteen years later. If shoreside parsons could volunteer 'for the duration', however long, then surely shoreside surgeons must also have done so, especially when the bloody encounters of 1588 were envisaged.

There were so many 'gentlemen volunteers' during the second and third Dutch Wars in the reign of Charles II that Pepys, the Secretary of the Admiralty, found it necessary to regulate them, since he could discern the need for greater professionalism in the Navy's methods. In the eighteenth century, also, volunteers went to sea, though some of them did so quite openly for the purposes of privateering under Letters of Marque in the capture of merchant ships belonging to enemy nations. Though the prime motive of the seagoing volunteer in every age and each national emergency was obviously to fight for sovereign and country, it must be that sheer love of adventure and the possibility of glory, honour and, of course, profit also played their part. The profit could come by way of prize-money, which was only abolished at the end of the Second World War. In 1762, for instance, the capture of the Spanish treasure-ship *Hermione* by two frigates produced for their crews £519,705, and the share-out varied from £65,000 to each captain and £485 to each seaman. However, all that apart, it must be conceded that few are not proud to say, concerning some engagement or campaign which is now enshrined in history, 'I was there.'

No. 34 First Company.

T. Hoath, Printer, Little Tower Street.

RIVER FENCIBLES

OF THE

CITY of LONDON.

WE HEREBY CERTIFY, That Edward Syer —

Twenty One Years of Age, Five Feet, Seven Inches high, Dark Complexion, Dark [illegible] Dark [illegible] is a Member of the [illegible] Corps, and that their Offer of Service has been a[illegible] of and accepted by [illegible] Majesty.

Captain Commandant.

Jno Drinkald Captain.

[illegible] 27. 1806

The Sea Fencibles

In the invasion emergency following the French Revolution, the Sea Fencibles were raised in 1798, consisting of volunteer fishermen and coastal seamen living all around the coasts. They reached a total of 30,000, and remained active until 1810. The River Fencibles, consisting of watermen and others connected with the River Thames, were founded in 1803 and disbanded in 1813, having seen action off Walcheren and at the capture of Flushing.

The Royal Trinity House Volunteer Artillery

At the height of the invasion scare in 1804, when Napoleon's 'army of England' was encamped at Boulogne, the Elder Brethren of Trinity House offered to equip, officer and man ten frigates then laid up for lack of crews, and to moor them across the Thames below Gravesend. At that time the Sea Fencibles were unpopular with the Navy because the men would not serve away from their own coasts, and Trinity House raised 1200 men in what was called the Royal Trinity House Volunteer Artillery, composed of merchant seamen, who drilled on Tower Wharf and in the open space which became Trinity Square. The first four frigates were manned in October 1803, and these were soon followed by the other six. When the news of Trafalgar was received late in 1805 the Trinity House Volunteers were disbanded.[3]

When it was realised that during 1811–13, when 29,405 men had been pressed into naval service by gangs employing 3,000 seamen, no less than 27,300 had later deserted, that system of

LEFT *Matthias Prime Lucas, Captain Commandant of the City of London River Fencibles 1803–1813 who subsequently became Lord Mayor of London in 1827. On 8 January 1806 at the River Funeral Procession of Lord Nelson, from Greenwich to Whitehall Stairs, the procession was attended by a considerable number of gunboats and rowboats of the London River Fencibles.* (*Guildhall Library*)

recruitment came to be called in question, especially on economic grounds.[4] Moreover, the big run-down in ships and men after 1815, and the consequent block in naval promotion, naturally curbed the opportunity for volunteers at a time when ten officers sought every vacant post. Thus, before long, the old Navy was dying, reform was in the air and the very earliest steps began to be contemplated which were ultimately to lay the foundation of the navy of Fisher and Jellicoe which was to come many decades later. One such step was actively to consider the formation of a corps of reserves which could be embodied into the Royal Navy as and when need arose.

The Royal Naval Coast Volunteers

In 1846 some public attention was drawn to the necessity of a naval reserve of fully-trained seamen, and naturally the Merchant Service was seen as the only direction in which it could be satisfied. In due course the Naval Estimates for 1852 contained provision for the enrolment of 5,000 such reservists. Though this clause was dropped almost immediately, the Royal Naval Coast Volunteers were set up by an Act passed on 15 August 1853, its members being drawn from seamen and others by voluntary enlistment, to the exclusion of deep-sea seamen and men who might seek regular service with the Royal Navy. Uniform was lent to them for their four weeks' training per year, during which time they received the pay of able seamen. Their training took place in naval ships but they were not to be carried more than fifty leagues (i.e. 150 miles) from the shore in peacetime or 100 leagues when called-up in wartime. The men who enlisted were largely inshore fishermen and their duties were mainly those involved in coast defence. At first these volunteers were under the control of the Coastguard Service, but from 1856 the Admiralty took over that responsibility, permitting Coastguard officers to command and train them. The Royal Commission of 1858 which enquired into manning of the Navy referred to the Coast Volunteers depreciatingly, as 'boatmen, fishermen and longshoremen', while Captain Cyprian Bridge, who had had them on board his ship in batches of 200 at a time, noted in his diary in 1863 that they were almost inevitably dirty, and Admiral Sir B. J. Sulivan wrote of 'shipping cabmen and others who had never been to sea'. Nevertheless a small number of Coast Volunteers had served under Sir Charles Napier in the Baltic during the Crimean War, despite the distance-limit placed on their service in wartime, which limit was not abolished officially until 1863. By that time, however, the (original) Royal Naval Reserve had come into being in 1859, with its insistence on 'active seafaring men only' to the exclusion of longshoremen. This new and important development promised a bleak future for the Coast Volunteers, and they were eventually disbanded in 1873.

The Royal Naval Artillery Volunteers

It had been unfortunate for the Coast Volunteers that they were brought to birth at a time when the 'evening of the old Navy' was beginning to turn into night, and before some of the hard lessons experienced during the Crimean War had been learnt. Officers and men serving in the 1850s and early 1860s had to come to terms with some sweeping changes in naval life. All new-building ships were steamers after 1851, and almost unbelievably the 'ship-of-the-line' itself was a steamer from 1860. For the senior officers, who had assimilated the old traditions in much less disturbed times, the changes must have been indigestible – hence the persistent requirement that seamen, including the yearned-for reserves, should be 'real sailors'. The Navy's hesitancy and unwillingness to make a proper job of the formation of a volunteer reserve of landsmen with a love of the sea violated the lessons of its own long history, and pointed to a touchy and unsure professionalism which was only beginning to develop its powers. Having excluded deep-sea seamen from the Coast Volunteers it had no justification for treating them with such contempt. Unfortunately the Navy's frame of mind concerning landsmen-volunteers in general was to continue for many years, and certainly during the time of its next venture into the same domain. In 1873, the year in which the Coast Volunteers were disbanded, an Act passed on 5 August set up the Royal Naval Artillery Volunteers, which was intended to provide trained gunners for service within the home seas and consisted for the most part of yacht-owners and professional men of good social standing.

The prime mover in the foundation of this latest reserve was Thomas Brassey, M.P. for Hastings, who himself was a keen amateur sailor and the founder of *Brassey's Naval Annual*. The First Lord of the Admiralty was sympathetic from the beginning, but the serving officers and men of the Navy cannot have enthused about this new opportunity for landsmen at a time when the favoured (original) Royal Naval Reserve, now fourteen years old, was not coming up to the expectations which had been held out for it. Even if the Navy took the new R.N.A.V. seriously comment could well be saved for the time when they made a hash of things. The R.N.A.V. never did that, despite their organisation into brigades and batteries in Army fashion, the rating of them into 'efficients' and 'non-efficients' rather than seamen, and their exclusion from an allotted place in national defence. Their officers' braid was silver and not gold, betokening an authority limited to assigned batteries only. Training was mainly in gunnery, small-arms, cutlass and boatwork. The Admiralty, however, did provide the London Brigade of the R.N.A.V. with an old gunboat, the *Rainbow*, which was replaced in 1888 by the composite screw gunvessel *Frolic*, both of which were moored in the Thames opposite Somerset House; and a yearly allowance of shot, shell and blanks. Beyond these somewhat grudging gifts, and until a small capitation was introduced, all other expenditure fell upon the pockets of the 'rowing and yachting gentlemen of good social standing' to whom the appeal had been made to set up the Volunteers in London, Brighton, the Clyde, Bristol and Liverpool. They paid an annual subscription and all their own expenses, bought their own uniform and ships' gear, and, in London, even hired a river-pier. The London men sailed down the Thames in their own boats, and on separate occasions manned first a schooner and then a yacht, both of which had

The R.N.A.V. drill-ship, the Rainbow, *moored upstream of Temple Steps.*

been provided by the more wealthy Volunteers, sailing them into the open sea. Their keenness and enthusiasm knew no bounds, and both discipline and maintenance were exemplary. The London Brigade of the R.N.A.V. was frequently in the public eye through its exuberance, and by its larger-than-life representation of the Navy it stole the show.

In the 1870s and early 1880s the nation was woefully indifferent about the Royal Navy, but in 1884 the first signs of a change in this attitude became apparent. Early in that year Russia invaded Afghanistan, and one of their generals went so far as to state that it was 'a political necessity for Russia to possess herself of India'. On 15 September a journalist, W. T. Stead, published in the *Pall Mall Gazette* the first of a series of sensational but informed articles entitled 'The Truth About the Navy' and described as from the pen of 'One Who Knows the Facts'. These articles were widely quoted, every newspaper and periodical began discussing the state of the Navy and the Government of the day (Gladstone's Second Ministry) became uneasy. The leading figure behind the articles was H. O.

Arnold-Forster, a well-known student of naval and military affairs and a good supporter of the voluntary spirit. He had not only suggested the articles but had also undertaken to provide for the *Gazette* the necessary information. In due course it became widely known that Commander John ('Jackie') Fisher, at that time the Commander of H.M.S. *Excellent*, was behind the *Pall Mall Gazette*'s exposures. Thus it was Fisher, together with Arnold-Forster and the Hon. Reginald Brett (later Lord Esher), a Machiavellian figure who always worked behind the scenes, who brought about the earliest of the three 'naval panics' of the 1880s and 1890s. The agitation in the Press during this and the second 'panic' of 1888 led to the big change of heart in the country concerning the Navy, and to the passing of the Naval Defence Act of 1889 which in time produced sixty-eight new warships.

Alas, this experience did not save the R.N.A.V. from dispersal, but it is difficult to avoid the conclusion that the R.N.A.V. who usually received a good Press even at a time when the Royal Navy was being scrutinised unfavourably, must have played some part in the revival. Their articulate spokesmen had no hesitation in contributing to the correspondence columns of newspapers, but this tendency appears to have assisted in their downfall, especially when disputes arose in public and resignations followed. It was a situation which irritated the Royal Navy, and the misunderstandings, frictions and jealousies which had long been simmering were finally considered by an Admiralty Committee under Vice-Admiral Sir George Tryon, which recommended that the Royal Naval Artillery Volunteers should be disbanded. Accordingly their end came on 1 April 1892, a few months prior to which date they amounted to sixty-six officers and no less than 1849 men. History had repeated itself, which is commonly accounted as tragedy. The London Brigade's subsequent claim for the reimbursement of £15,272 which it had spent on the *Frolic* and associated expenditure was officially cut down to £5,939. Though the Navy had little time for the Volunteers they are assured of immortality through Lieutenant C. R. Low's respectful inclusion in his *Her Majesty's Navy* of 1890–3 of two fine chromolithographs by W. Christian Symons which show men of the R.N.A.V. beside the guns of the Fleet.

Though the Royal Naval Artillery Volunteers had ceased to function, the spirit of service which had motivated them lived on to find expression in cruising clubs in several areas where the Volunteers had been active, but these were a poor substitute for what had been lost. The disbanded diehards formed a National Resuscitation Committee which took every opportunity to publicise their sorry state and to lobby all who were in a position to bring influence to bear upon it. However, as the years passed fresh champions emerged, and by 1900 a new National Committee, led by the Marquess of Graham (later the Duke of Montrose) and C. E. H. Chadwyck Healey, Q.C., who had been an officer in the R.N.A.V., and including survivors from the old Resuscitation Committee, began to operate powerfully, having originated through a crowded meeting held under the chairmanship of the Lord Provost of Glasgow on the evening of Trafalgar Day, 21 October 1899. Graham and others persuaded the then Lord Mayor of London to convene a meeting at the Mansion House, which was supported by eight Lord Mayors, all the Lord Provosts of Scotland, over 100 Mayors and Provosts of towns and a similar number of Members of Parliament.

A Royal Naval Artillery Volunteer.

Certainly the time was ripe for a forthright assault to be made against the old prejudices which had killed the R.N.A.V. The Royal Naval Exhibition held at Chelsea in 1891 under the patronage of the Queen had been visited by well over two million people; the Navy Records Society, for the printing of documents and papers connected with naval history, biography and archaeology, had been founded in 1893; and the Navy League, mainly civilian in its constitution and pledged to securing naval efficiency adequate to Britain's world-wide imperial responsibilities, had followed in 1894. In addition, the Naval Reviews of 1887 (Queen Victoria's Jubilee), 1889 (which the Kaiser attended with a detachment of his fleet) and 1897 (Queen Victoria's Diamond Jubilee) had whipped up much popular fervour for naval matters, and amid remarkable patriotic feeling the Second Boer War had been painfully fought and won. In organising the passage of and armada of troopships to Table Bay in the autumn of 1899 the Royal Navy had performed a task the like of which it had not been given since it sent sailing ships to carry Lord Raglan and his 30,000 men to Scutari in 1854 at the beginning of the Crimean War. However, more powerful in influence on naval thinking than any of these events or circumstances was the growing naval might of Germany and her bid for supremacy over the British Fleet.

The omens came good officially for the volunteer cause in 1900 when Lord Selborne, a young peer who recognised the genius that was in Fisher, became First Lord of the Admiralty, and Arnold-Forster, the first stirrer of the agitation over naval affairs in 1884 and a joint-architect with Fisher of the 'naval panic' of that year, was appointed Parliamentary Secretary to the Admiralty. Graham's new National Committee and its large number of influential supporters drew up an appeal to Lord Salisbury, who had been Prime Minister since 1895 and who had just been re-elected, and Graham had a meeting with him and Lord Selborne. At about the same time, Arnold-Forster sent for a former member of the R.N.A.V., H. Warington Smyth, the Honorary Secretary for London of the National Committee, who still smarted over the disbandment, and informed him of a change of heart concerning naval volunteers that had taken place within the Admiralty under Lord Selborne's leadership. One of the questions put to Smyth and about which he was to provide the answers was crucial – would the projected Volunteers be willing to serve anywhere in the world, and to perform all duties required on board ship? When Lord Selborne spoke in the House of Lords in July 1901 he made the same point, saying it was 'imperative', adding that this was 'thoroughly understood by those gentlemen who are interested in the scheme'. They could not have wished for more, and the new Volunteers were obviously going to be of a different species in terms of service from the old R.N.A.V. Warington Smyth, already an experienced yachtsman and traveller in South-East Asia, who was later to have a distinguished imperial career, had given Arnold-Forster the necessary assurance.

In 1902, after Graham and others of the National Committee had held discussions presided over by Admiral Sir Gerard Noel at the Admiralty, the matter was remitted by the Admiralty to a new Naval Reserves Committee under Sir Edward Grey, M.P., in order to frame the general lines on which a proposed Volunteer force would operate, and Graham was appointed a member of it. In due course Sir Edward Grey was able to report to the House of Commons on the work which this committee had done, and when he presented the Naval Estimates in 1903 he said, 'It seems to be both wasteful and unnatural that all the amateur sea talent in this country should for want of opportunity be obliged to turn to military service to the exclusion of Naval training'. He then announced that the Admiralty would establish the Royal Naval Volunteer Reserve. The title was Chadwyck Healey's idea, finally preferred by Sir Edward Grey after consideration of two other submissions from within the committee – Imperial Naval Volunteers and Naval Imperial Volunteers[5]. A force of Royal Marine Volunteers was also to be set up.

By tradition, the supervisor of all matters relating to the Navy's personnel, including training, is the Second Sea Lord, an office which, before 1904 and for about a century, was styled Second Naval Lord. At the vital moment, from the point of view of those who were pressing for the new Volunteer Reserve, the Second Naval Lord was none other than the compulsive reformer Fisher, who held that office from June 1902 until August 1903. Only two months before he left it, to become briefly Commander-in-Chief at Portsmouth, Parliament passed the Naval Forces Act (1903), which authorised the Admiralty to 'raise and maintain a force to be called the Royal Naval Volunteer Reserve'. Thus the R.N.V.R. was born at a time when Fisher's massive plans for reforming all branches of the Navy were beginning to take effect through the 'Selborne Scheme', which was made known at Christmas 1902 and introduced in September 1903, and was able to take some of its earliest steps at the same time as the Navy itself was being reborn, under Fisher as First Sea Lord from 1904 to 1910. It was signal and poetic justice.

In little more than ten years from the date on which the R.N.A.V. was put to death by unthinking prejudice – a sin which calls to mind the burning of Joan of Arc – the Royal Navy, with strong men now at the helm, created, at its third attempt, and by using men of the same calibre as the Artillery Volunteers, a new Reserve capable of mobilisation in wartime for service anywhere in the world. In 1945, towards the end of the Second World War, the R.N.V.R. was providing 74% of all the executive officers then serving in the Royal Navy, although, of course, the majority of these were 'Temporaries' serving only for the duration of hostilities, and not permanent pre-war members of the several R.N.V.R. Divisions who had been mobilised in 1939. That a chosen few were actually in command of H.M. Ships must have caused consternation in the Victorian section of Valhalla, with, here and there, wry smiles among the Artillery Volunteers. In 1903, however, such a consummation, even if it was thinkable, would have seemed a romantic dream quite beyond the bounds of reality, because, for some years to come, the new Reserve was to suffer, like even Fisher and his reforms, from some residual dogmatic prejudices which only bitter wartime experience was to remove.

CHAPTER TWO

'The Blackfriars Buccaneers'

Following the passing of the Naval Forces Act on 30 June 1903 the Admiralty wasted no time in implementing it. The first two appointments of Commanding Officers of the new R.N.V.R. Divisions appeared in the *London Gazette* of 7 August – the Hon. Rupert Guinness (later Earl of Iveagh) for London Division and the Marquess of Graham for Clyde Division, each as Lieutenant Commanding. In the matter of naval seniority, as 'Gr' preceded 'Gu', it was not Guinness but Graham who received the No. 1 R.N.V.R. Commission, as Graham explained in 1952 in his autobiography *My Ditty Box*. It was fair that he should receive the honour for he had worked very hard to establish the new Reserve. Both were personal friends of Fisher, who at one time hoped that Guinness might marry one of his daughters.

The regulations and administrative rules for the constitution of the R.N.V.R. were issued in an Admiralty Blue-book on 10 August[1]. Several old ghosts were laid in the overall provision that whenever the Admiralty called out the Volunteers, or any of them, for actual service they were liable to do any duty applicable to their rating for which they were required by the Commanding Officer of the ship or force to which they were attached. They would be subject to the Naval Discipline Act and they would accommodate themselves to the berthing and messing arrangements usual for the seamen of the Royal Navy.

For administrative purposes the Volunteers would be formed into Divisions, each of not less than five numbered Companies of 100 men. Each Division would be commanded by a Lieutenant Commanding, and each Company would be commanded by a Sub-Lieutenant. Each Division would have twenty-six Commissioned Officers, thirty-two Non-Commissioned Officers and 465 men, making 523 in each Division. Included in the Commissioned Officers were to be a Staff Surgeon (Honorary), five Surgeons, a Chaplain (Honorary), two Paymasters and five Midshipmen. A number of trained Signalmen were to be added to the permitted establishment as the Admiralty would direct. Among the Leading Men and Men there should be as many as possible who were qualified by trade to act as telegraphists, electricians, armourers or artisans. Lieutenant Instructors, Petty Officer Instructors and Armourers were to form the Permanent Staff, and would be appointed by the Admiralty.

The whole new force was to be administered by a central Admiralty Volunteer Committee, comprising representatives of the local Divisions, a Naval Officer representing the Admiralty, and 'other gentlemen' whom the Admiralty would invite. Office accommodation and contingent expenses therefor would be provided by the Admiralty, and the nationwide R.N.V.R. would be under the orders of the Admiral Commanding Coast Guard and Reserves. All R.N.V.R. officers would be appointed by the Admiralty, and would receive Commissions in the same manner as Royal Navy officers. When called out for War Service the Officers and Men of the R.N.V.R. would receive the continuous service pay of the Royal Navy appropriate to their rank or rating, and if injured, killed or drowned on duty comparable compensation for themselves or pensions and allowances for their wives and children.

Uniform worn by officers of the R.N.V.R. was to be of the same pattern as for Royal Navy officers, but with certain excep-

Commander The Hon Rupert Guinness, First Commanding Officer of London Division.

tions – such as *waved* quarter-inch gold braid instead of the usual half-inch gold lace on the sleeve, or buttons to which the letters 'R.N.V.' had been added in old English characters. Each Naval Volunteer was to receive free of cost from the Admiralty the kit appropriate to his rank, but again with certain exceptions – such as buttons worn by chief petty officers were to be the same as those worn by their officers, and those worn by petty officers were to be the same pattern but of black horn, while the three white tape trimmings on the 'sailor's collar' were to be of parallel *waved* lines, and cap ribbons were to bear the letters 'R.N.V.' in Roman lettering. No deviations from these established patterns were allowed.

Such were the chief details on the birth certificate of the R.N.V.R., which in time and until it was merged with the existing R.N.R. in 1958 became known colloquially as the 'Wavy Navy'. The Regulations of 1903 must have been thought unbelievably generous by those who had known the R.N.A.V. Gone now were the 'longshoremen' view of volunteers, with its Ruritanian requirement for silver braid to be worn by officers, the military symbols of brigades and batteries, the demeaning limitation of officers' authority to their own batteries and the

No 1 Company formed from the Stock Exchange by Lieut. Edmund Wildy, R.N.V.R.

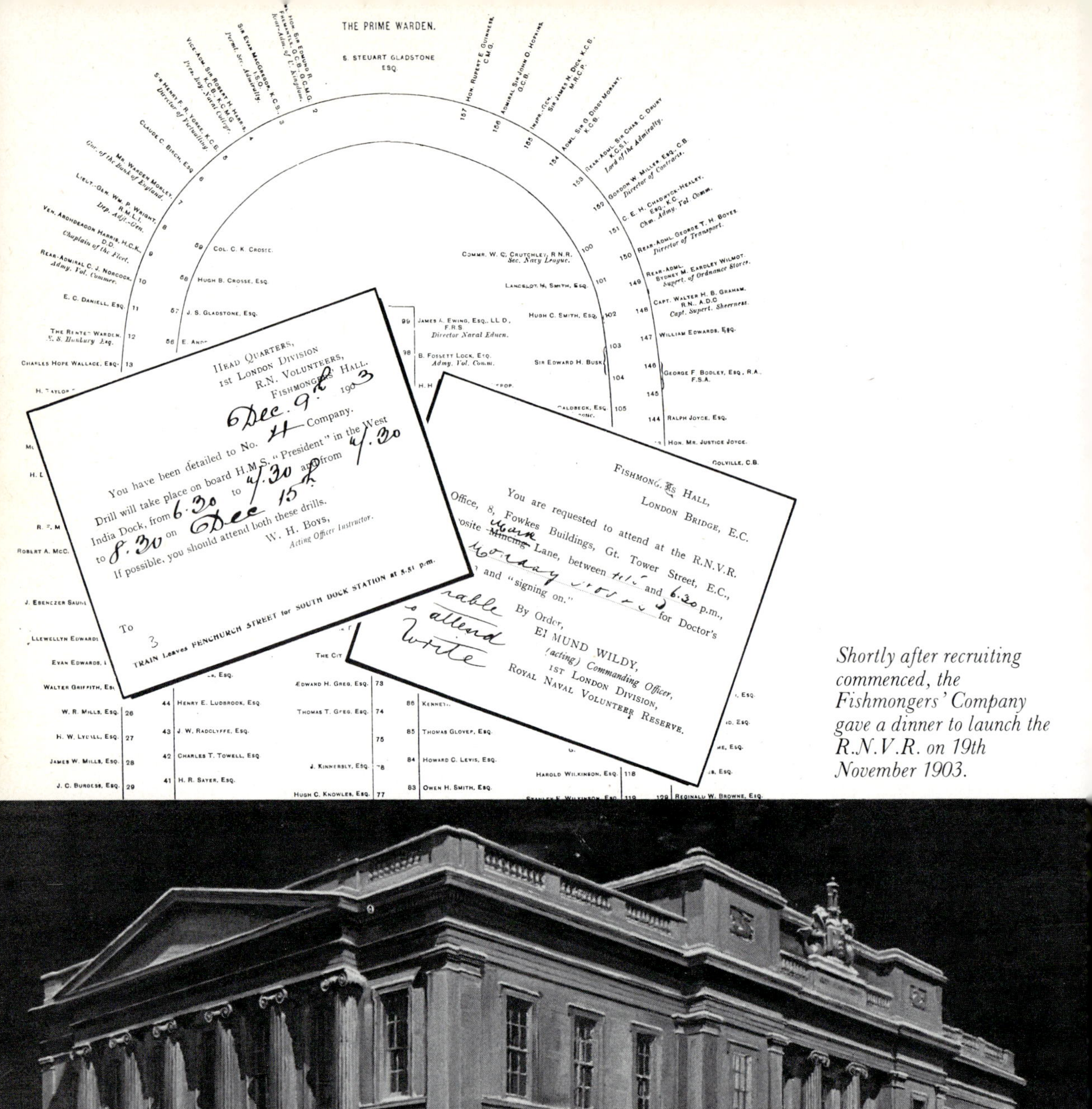

HEAD QUARTERS,
1st London Division
R.N. Volunteers,
Fishmongers' Hall.
6 Dec. 9th 1903

You have been detailed to No. 4 Company.
Drill will take place on board H.M.S. "President" in the West India Dock, from 6.30 to 4.30 p.m. and from 4.30 to 8.30 on Dec 15th.
If possible, you should attend both these drills.

W. H. Boys,
Acting Officer Instructor.

To 3

TRAIN leaves FENCHURCH STREET for SOUTH DOCK STATION at 5.51 p.m.

Fishmongers' Hall,
London Bridge, E.C.

You are requested to attend at the R.N.V.R. Office, 8, Fowkes Buildings, Gt. Tower Street, E.C., opposite ~~Mincing~~ Mark Lane, between 4.15 and 6.30 p.m., Monday Nov 23 for Doctor's ... and "signing on."

By Order,
Edmund Wildy,
(acting) Commanding Officer,
1st London Division,
Royal Naval Volunteer Reserve.

Shortly after recruiting commenced, the Fishmongers' Company gave a dinner to launch the R.N.V.R. on 19th November 1903.

possibility that the social status of the Volunteers would cause difficulties on the lower deck during service with the Fleet. While the fresh uniform differences which had been laid down were accepted, and even became a matter of pride producing strong protest when they were abolished in 1951, the Admiralty's unwillingness in 1903 to concede to the Men of the R.N.V.R. the designation of Seamen was not suffered lightly. It was clear that the R.N.V.R. would have to win its way to true brotherhood with the Royal Navy, but perhaps the decision was wise in that it provided a spur to efficiency. Rectification of this was to take six years.

Recruiting for the new London Division began early in September 1903, and by the end of the month Chief Petty Officer Goodyear had enrolled over 700 recruits, while he sat daily, including Saturdays, at Fishmongers' Hall, London Bridge, from 10 a.m. to 5 p.m. A good sprinkling of professional City men had been among the applicants. It was an excellent response considering that the men came from all parts of suburban London and for the time being would be expected to attend on board the Royal Naval Reserve's old wooden sloop *President* (formerly the *Gannet*) lying in West India Docks at Millwall, which necessitated a train journey from Fenchurch Street Station after office hours. The first R.N.V.R. drill on board her took place on 10 November.

All recruits had to be British subjects, over eighteen, physically fit and prepared to serve for three years, and it was made known that men only five feet tall with no chests to be proud of were as welcome as men six feet tall with fifty-two inch chests. Maybe it was this which caused a story to go round that when Surgeon F.J. Hannan was medically examining recruits in a small dingy office in Fenchurch Street girls working in an office across a ventilating-well requested that the blinds should be pulled down.

It had been arranged that London Division was to have as its new drill-ship the composite screw sloop *Buzzard* of 1140 tons which was to lie in the Thames off the Embankment slightly upstream from Blackfriars Bridge, but she was at Chatham Dockyard being prepared for her new role and was not yet ready. Eventually she was sent to Sheerness for the removal of her engines and boilers and after some delay was towed up the Thames by three tugs to her permanent mooring, passing under Blackfriars Bridge in the late afternoon of 9 May 1904, without her funnel and rigging and looking bare and gaunt.

The name *Buzzard* had to be retained for her by the Division because the R.N.R. drill-ship *Gannet* had been renamed *President* on 16 May 1903, just before the R.N.V.R. was born, and was in service until 1911 when she was lent to Mr. C. B. Fry for use ultimately as the Training Ship *Mercury*. This disposal enabled the *Buzzard* to be renamed *President* on 1 April 1911 as the 7th

President *(in background) in West India Dock where the first drills took place in 1903.*

H.M. Ship to bear the name, which was originally bestowed on a naval ship in 1645 in honour of the then Lord President of the Privy Council.

The *Buzzard* began her new R.N.V.R. career on 18 June 1904 when she hoisted the White Ensign at 8 a.m. Her new Commanding Officer, Commander the Hon. Rupert Guinness, who had been promoted to that rank on 7 March, the Instructor, Commander Hewetson, R.N., and certain other officers had gone on board the previous evening to take possession of the ship. The *Buzzard* was declared open for training on Saturday, 2 July, when her R.N.V.R. crew were inspected by the Lord Mayor and Lady Mayoress, the ship being dressed overall for the occasion. An 'immense crowd' watched the operations from the Embankment, an exhibition of gun-drill with the shielded 5-inch and 6-inch guns on the main deck and their crews of nine and ten men each being the chief attraction. Selected targets on or near the river, including the Houses of Parliament and the offices of the Metropolitan Asylums Board, of which Guinness was Chairman, were 'bombarded'. Among the distinguished visitors present that day were Lord Selborne and Lord Brassey, who were worthy representatives of all those who had championed the Volunteer cause. A general muster of the men was held in the afternoon in St. James's Park, whence they marched in fours to the Embankment and halted near the landing steps of the ship. The Guard of Honour was 100 strong and was under the command of Lieutenant Edmund Wildy, R.N.V.R.

Commander Rupert Guinness taking over his drill ship Buzzard *on 17 June 1904. The proceedings were devoid of ceremony. Stepping from the steam launch he had boarded at Temple Steps he went over the* Buzzard*'s side followed by his Officers each reporting as he did so, 'Permission to come on board, Sir.'*

In addition to the drill ship *Buzzard* the Division soon acquired a large building, hitherto used for the storage of steel, at Princes Wharf, Commercial Road, Lambeth, on the south side of the river opposite Temple Pier, as a drill hall in which the whole of the Division could be drilled at one time. The rental of this was said to be £600 per annum, of which Commander Guinness made himself responsible for £500. Surprisingly, the old 'patronage' and 'self-help' concepts which had marred the R.N.A.V. still encumbered the scene even in the heady days of 1904.

Whereas the men of the Division had received their uniform (boots and heavy weather gear excepted) at the charge of the Admiralty, the officers were required to buy their own uniforms, and three years were to pass before an Admiralty grant of £20 towards this expense was authorised. Travelling expenses to and from the forty drills in the first year and twenty-four in each subsequent year had to be borne personally by both officers and

HMS Buzzard.

Lieut. Edmund Wildy (on board Buzzard*) was the first Executive Officer of London Division and in command of No 1 Company.*

men, and the R.N.V.R. was expected to maintain its headquarters and to pay for all lighting, heating and repairs. Officers and men gained pride from their membership of the R.N.V.R. and they were prepared to pay for it, which standard was possibly set by those who had formerly served in the R.N.A.V. To some degree this may be seen as a natural reaction to the Admiralty's disinclination to class them as seamen; if they were compelled to be different in that respect they were willing to be different in paying their way, thus demonstrating that so far as patriotism went they were second to none.

As part of a fledgling Reserve, London Division had to ensure there was no going back to the self-assertive ways of the R.N.A.V. which had proved an embarrassment to the Royal Navy and a godsend to the cartoonists. The Navy itself had changed since those days, and, with Fisher about to go to the highest office, would change still more. Nevertheless, Fleet Street was not willing to be deprived of its handy nautical victims, and, to the embarrassment now of the Division, continued to comment banteringly in the old tradition, dubbing its members: 'the Blackfriars Buccaneers' and 'the *Buzzard* Boys'. If the R.N.V.R. suffered any persisting disabilities under the Admiralty's qualified blessing on it, the only hope could be that, as its reputation grew, it would earn their removal by its keenness and efficiency.

As far as numbers were concerned London Division led the way with about 1000 men, in ten Companies. Though there was the seamen's work of compass and helm, heaving the lead, small boat work with cutters and pinnaces, knots and splicing, and making up a hammock from mere canvas and rope, gunnery was the principal task. The *Buzzard* could show examples of many of the Navy's guns – 6-inch, 5-inch, 4.7-inch, 12-pdr., 6-pdr., Maxim and field guns, as well as Admiral Scott's 'dotter' for target practice and loading machines for teaching quickness in working a gun's crew[2]. It was also possible for a Volunteer to specialise in signalling or, if he was interested in engineering, on work with the steam cutter. Much of the foregoing could be carried out in the Commercial Road drill hall, but on board the *Buzzard* each Company in turn formed the ship's company. Each man in the Division could live on board the ship for a fortnight every twenty weeks. The ship was kept like a seagoing vessel, but the crew could only go on board in their spare time, which meant after their day's work in the City office or workshop was finished.

SCRUBBING DOWN DECKS BEFORE LEAVING FOR THE CITY

The volunteers in the Buzzard, who are workers in the City, go on board in the evening and return to the City in the morning in time to resume work. In the picture some of the men are seen scrubbing decks before breakfast, after which they resume their civilian clothes and go to work.

Cutting from Daily Mirror *15 September 1908.*

On a typical evening the men began to arrive at about 5 p.m., and soon boats' crews were formed. The boats were then lowered, manned and sent away, while other men carried on as guns' crews or attended lectures. At 8 p.m. all hands were piped to supper, which was served on the lower or mess deck, and at 9 p.m. to 'Stand by hammocks', when each man obtained his hammock from its place in the hammock-nettings and slung it ready for turning-in on the lower deck. During this, 'Fire quarters' or 'Close all water-tight doors' or 'Out collision mats' could be piped, when everyone left his work in hand and manned his proper station. Afterwards 'Rounds' would be piped, when the crew formed up on deck, all lights were doused, and the Duty Officer went round with the quartermaster to ensure all was secure for the night. This done, the crew were at liberty to turn into their hammocks, though sometimes an impromptu concert was raised until 10 p.m., when all had to retire.

At 6 a.m. next morning came the pipe to 'Lash up and stow', when all tumbled out, lashed up their hammocks, stowed them in their proper places, and fell in for cleaning ship. This was carried on until 7.45 a.m. when, all gear being cleared away, hands were piped to get into their 'shore rig', and, all having breakfasted, the first boat left for the shore and the day's work. At 5 or 6 p.m. that evening the men were back again for another night's duty.

The Admiral Commanding Coast Guard and Reserves, Admiral Rice, who had been a member of the Tryon Committee which disbanded the R.N.A.V. in 1892, conducted the Division's first Annual Inspection in December 1904. He was received by Commander Guinness, Commander Hewetson, R.N., Paymaster Greenwood, Sub-Lieutenant Hampshier, Mr. Chadwyck Healey, K.C., who was the Chairman of the Admiralty Volunteer Committee, Mr. Lock and Chief Gunner Gunn. He expressed his pleasure at the smartness and efficiency

Sailors stowing their hammocks on board Buzzard *before working 'part of ship'.*

A Company preparing for Admiral's Inspection on board HMS Buzzard *on the 21 October 1905, the 100th anniversary of the Battle of Trafalgar.*

shown, and went on to say that the appearance and bearing of the Guard of forty men from No. 6 Company was such that he found it difficult to realise that he was not stepping on board one of His Majesty's Ships in commission. Later he inspected the remaining nine Companies of the Division in the drill hall south of the river, and said he had never seen a finer body of men than those he was addressing, expressing the hope that the City of London would show some practical token of its appreciation. He promised to give a good account of the Division to the Admiralty when he reported. From Admiral Rice it was praise indeed. In the following two years Vice-Admiral Henderson, Rice's successor, expressed his satisfaction in similar terms.

Apart from the routine training which took place on board the *Buzzard* and in the drill hall some training was done farther afield, near Brighton in 1904 and near Eastbourne in 1905 and 1906, each during the Easter period. Later in 1905 the first personal contact was made with H.M.S. *Excellent* at Portsmouth, when a group of C.P.O.'s from all R.N.V.R. Divisions assembled there for a short course of gunnery instruction. At about this time also signalmen from the Division attended a special course of instruction. The required brotherhood with the men of the Royal Navy was slowly growing.

A specially interesting new note was struck in the summer of 1905 when a contingent of London Division joined the Reserve Fleet battleship *Barfleur*, which possessed only a 'nucleus' crew and was acting as Guardship to the King at Cowes Regatta, thereby assisting in bringing her up to full complement for that purpose. Others from the Division went at the same time to the cruiser *King Alfred* at Portsmouth, where they found 'Lash up and stow' was piped at 5 a.m. Perhaps on the basis of this there was some attempt by the Press to imply that the Division's City men had found the 'spit and polish' on board these ships irksome, but one journalist who had tried to develop the idea had to concede that a man had said to him, 'Though at times I would catch the officer grinning at us I believe they learned to respect the Volunteers'. Yet another journalist said that a City clerk had told him, 'Every one will go again next year.' It was by such attitudes that the men of the R.N.V.R. built up their pre-war reputation.

Despite what had undoubtedly been achieved, a rumour spread in the summer of 1907 that the R.N.V.R. was going to be disbanded, and a question on the matter was actually asked in Parliament, to receive a complete denial. The rumour was possibly connected with a report earlier in that year that the Division had 350 vacancies in the permitted maximum strength of 1000, owing to the resignation of some of the men who had been the first to join on the formation of the Division. Maybe in an endeavour to boost recruitment Admiral Lord Charles Beresford, ever a friend of the R.N.V.R., who was in command of the Channel Fleet, saw to it that a large number of men of the R.N.V.R. accompanied its Spring Cruise in 1907. Before they finally went ashore at Dover the Admiral called them all on board his flagship, the *King Edward VII*, in accordance with his custom and told them he was very pleased with the excellent reports about them which he had received from the captains of the ships in which they had been borne. Nevertheless there was some adverse criticism, as might be expected when the Navy still contained many who could never come to concede that landsmen could be made into sailors. The Captain of the cruiser *Bacchante* wrote: 'The result has been most satisfactory; nothing

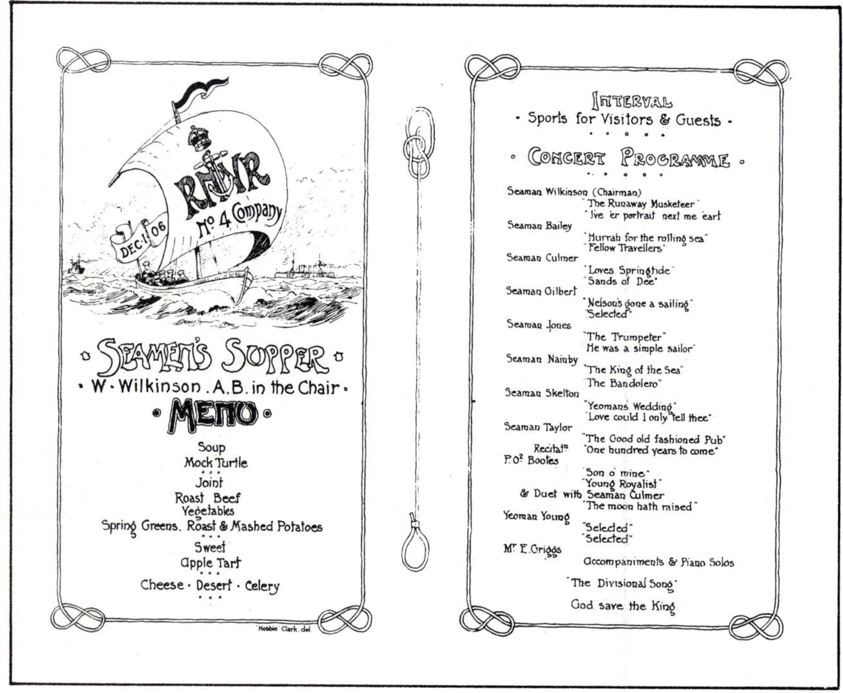
RNVR
No 4 Company
DEC 1 06

SEAMEN'S SUPPER
W. Wilkinson. A.B. in the Chair
MENU

Soup
Mock Turtle
Joint
Roast Beef
Vegetables
Spring Greens. Roast & Mashed Potatoes
Sweet
Apple Tart
Cheese · Desert · Celery

'Nobbie' Clark. del

INTERVAL
Sports for Visitors & Guests

CONCERT PROGRAMME

Seaman Wilkinson (Chairman) "The Runaway Musketeer" "Ive 'er portrait next me 'eart"
Seaman Bailey "Hurrah for the rolling sea" "Fellow Travellers"
Seaman Culmer "Loves Springtide" "Sands of Dee"
Seaman Gilbert "Nelson's gone a sailing" "Selected"
Seaman Jones "The Trumpeter" "He was a simple sailor"
Seaman Nainby "The King of the Sea" "The Bandolero"
Seaman Skelton "Yeomans Wedding" "Love could I only tell thee"
Seaman Taylor "The Good old fashioned Pub"
Recitatn "One hundred years to come"
P.O? Bootes "Son o' mine" "Young Royalist"
& Duet with Seaman Culmer "The moon hath raised"
Yeoman Young "Selected" "Selected"
Mr E. Griggs Accompaniments & Piano Solos
"The Divisional Song"
God save the King

could exceed the keenness to learn displayed by both officers and men alike, and I am of the opinion that a few more trainings to such men will make them a valuable addition to His Majesty's Service. They take themselves seriously and must be taken seriously or the whole scheme will collapse.'

As the Admiralty slowly discovered, the annual cruises of R.N.V.R. men in battleships became the great selling-point in recruitment for the R.N.V.R. as the years went by. After all, it was a fine thing for a Briton to feel that he was qualified to take his place in his country's Fleet, and to serve under the White Ensign. By their patriotism, keenness and ability the men of the R.N.V.R. had abundantly earned the consideration which they increasingly received in Whitehall. In June 1908 all Divisions

Visit of the Prince of Wales in February 1907 at Commercial Road, Lambeth.

were invited to send ten officers and 126 men to form part of the escort which would accompany the then Prince of Wales (later King George V) on a forthcoming visit to Canada. Two officers, Lieutenant Ginman and Sub-Lieutenant McCormick Goodhart, and twenty-seven men went from London Division, the men having been chosen from among those who had made the most progress during the previous twelve months.[3] To participate in the visit, which lasted from 29 June to 18 August, was seen as a great honour for the whole R.N.V.R. Before 1914 the men of the Division were classed as 'efficients' and 'non-efficients', and only the former could be considered for service with the Fleet. Selection for the Canadian visit represented the very apotheosis of that principle. In London Division the Instructor, Commander Hewetson, R.N., laid a firm foundation for the training programme in pre-1914 days, and a written lower deck opinion of him still survives in the recollections of C. P. O. McDonell, a leading figure on board the *Buzzard*. He said: 'Cdr. Hewetson was a very good appointment. He tried in every way to understand us and was a gentleman in all his dealings.' The Division was fortunate in that it possessed fourteen officers junior to Commander Guinness who, having joined it in 1903 or 1904, were still with it when mobilisation took place in 1914. In the intervening years most of them had been promoted, and the *Navy List* for January 1915 shows them as still belonging to the Division, though of course serving elsewhere with the Royal Navy. These were: Lieutenant-Commanders E. Wildy, T. H. Roberts-Wray, and H. D. King; Lieutenants S. Searle and W. Ginman; Staff Surgeons A. R. Brailey, F. W. Smith and A. D. Cowburn; Surgeons F. J. Hannan, R. J. E. Hanson, E. J. Steegmann and C. D. Marshall; Staff Paymaster P. Nisbet; Paymaster C. Greenwood. The Bishop of London, Dr. A. F. Winnington-Ingram, remained as (Honorary) Chaplain and Sir Frederick Treves as (Honorary) Staff Surgeon, both of whom had served from the beginning. As ever, continuity had proved the greatest ally for success.

Between the spring and autumn of 1908 the Division was able to plan for contingents of its members to take part in five cruises with the Fleet. With the object of further familiarising them with the terms in use in the Royal Navy the old description 'Company' was dropped and 'Division' was substituted, the ten Divisions which resulted being known as Fo'c's'lemen, Foretopmen, Maintopmen, Quarter-deckmen and Mizzentopmen, with each of these split into Port and Starboard Watches. It was another step in the right direction.

An even greater step followed soon after. In the new Regulations for the R.N.V.R. which were published in 1909 the Admiralty at last conceded that the members were to be classed as Seamen. It was a great triumph, and it had been well earned. Instead of Leading Men and Men there were now to be Leading

Seamen, Able Seamen and Ordinary Seamen, as in the Royal Navy. The last troublesome vestige of the R.N.A.V. had been swept from the scene, and with it any idea that might have lingered in Fleet Street that the members of the R.N.V.R. were to be seen as figures of fun. In 1909 the wisest heads in control of the Fighting Services were certain that Britain was on the road to war, and many old naval prejudices were acknowledged to be unrealistic and inconvenient in the new situation. It was around this time that Fisher, who remained as First Sea Lord until January 1910, made his astonishing prediction, based largely on his calculation about the time Germany would take to widen the Kiel Canal, that the war would break out on a Bank Holiday in 1914 – which it almost did, on 4 August, the day after the August Bank Holiday.[4]

A special feature of the early days was the interdivisional field-gun competition held annually in Hyde Park near Knightsbridge Barracks. A press cutting of 4 October 1908 says: 'The men marched from their depot, Prince's Wharf, Lambeth, headed by their Band, causing through their smart appearance considerable interest en route'. The competition was for the smartest gun's crew in the Division. The trophy, presented by Commander Viscount Curzon and known as the H.M.S. *Terrible* Trophy, was a silver model of one of the 4.7-inch naval guns landed in South Africa for the Ladysmith Relief Column. The judges were R.N. officers from H.M.S. *Excellent*, the Gunnery School at Whale Island, and considerable crowds gathered to watch the event.

Such is the outline of the story of the first years of London

Negotiating the imitation wall during the inter-company Field Gun competition held annually in Hyde Park.

No 1 Company, winner of HMS Terrible Trophy in 1907 photographed at Lambeth Headquarters.

Division, as it arose, developed and then stood waiting for the call to arms. History, however, involves more than consideration of policy, performance and events, necessary though the recital of these things may be. At the same time, it is worth noting the actions and reactions of those who were actively involved, as they carried out their duties and gave of their time. Some small insight into this is provided, firstly, by correspondence which still exists from Lieut. T. H. Roberts-Wray's No. 3 Division and deals with its day-to-day matters which arose during the years immediately preceding 1914. This reveals a wide variety of letters to their Divisional Officer from members of the Division stating their problems, the majority of which relate to inability to put in the required number of drills due to change of job or of hours of work. Each request was dealt with on its merits, with or without the fine that could be imposed.

One Seaman penned a beautiful letter from New Brunswick, Canada, explaining how, as he had been 'without a berth' for two months, he had signed on as an Ordinary Seaman in the Elder Dempster ship *Kaduna* the day after he had attended drill on board the *Buzzard* one month previously. He realised the difficulty he had caused the Division and sought guidance 'how to go on', saying that if his kit was required to be returned his mother would forward it to the *Buzzard* 'carriage-paid'. He had written on the day after his arrival in Canada a fortnight overdue and after a rough passage, and he said that his ship would be going on to South Africa. The Division's reply has not been preserved. Another Seaman wrote saying that as he had not been in a good job in London he had been living with his parents at the H.Q. of the London Fire Brigade, from which his father, a fireman, was about to retire and to move to Devon. In the circumstances he said he was also moving to Devon experi-

Site of drill hall in Lambeth.

mentally and hoped to obtain work there, so he sought six months' leave from the R.N.V.R. at the expiry of which, if the experiment was successful, he would write again seeking his discharge. In this case, too, the reply has not been preserved. In yet another letter a married man with five children and an ailing wife sought his discharge saying he worked daily from 9 a.m. to 9.30 p.m. (11.30 p.m. on Saturdays) for thirty shillings a week, and added 'my children must come first'. Probably he was given his discharge, it is to be hoped without fine. Financial hardship was present in all these cases, and surely had to be accepted.

Of a more routine kind is C.P.O. Stanley Geary's letter of 1911 to his Divisional Officer listing fifteen non-efficient Seamen whom he recommended should be 'forthwith discharged'. Of these fifteen, the four who had been absent without permission from the Admiral's Inspection were specially noted. Geary went on in 1914 to obtain a Commission in the Collingwood Battalion of the Royal Naval Division, which was totally destroyed at Gallipoli on 4 June 1915, when twenty-four of its officers and 500 of its men became casualties. Geary, who was wounded two days later, was one of the few Collingwood Battalion officers who remained alive. After the war he was a leading figure in the Royal Naval Division Association. When he died aged eighty-seven in 1967 and was buried in Bandon Hill Cemetery at Croydon his war medals were resting on his coffin glinting in the sun and were buried with him.

When Able Seaman Kortright wrote to his Divisional Officer on 21 July 1910 with the information that his rifle had been missing since 19 May he was grasping a nettle which must have troubled him for two months. On that date he had taken his rifle to the drill hall to practise for King Edward VII's funeral. When he was not selected he put it down to watch others drilling and later found that it had been removed, another having been put in its place. He handed this in, asking for it to be given to its proper owner. A week later he was told an instructor had claimed it, saying that it had been taken from his store. Kortright could only end his letter with the sad words: 'I have heard nothing more of my rifle'.

There is a reassuring warmth about Lieut. Roberts-Wray's order of 21 January 1914 to the Petty Officers of his Division. So as to ensure that they were kept more in touch with individual men serving under them, three Petty Officers were given special duties with Seamanship Instruction, Gunnery Classes and the Divisional Whaler, while two Leading Seamen were made generally responsible for hunting up men who had not yet passed for Able Seaman or Leading Seaman, and for seeing that men attended the rifle range more regularly and improved their shooting skill. The Gunnery responsibility was given to P.O. L. L. Withrington, a burly man who became commissioned and who rose to be a Captain, R.N.V.R., though he never commanded London Division. In August 1940 when Withrington was briefly a patient at the Royal Naval Hospital at Chatham, high up on the Lines above the town, the author of this study recalls walking with him in the hospital garden in brilliant sunshine while the vapour-trails of Spitfires fighting the Battle of Britain crossed the very blue sky above and he talked about life in the *Buzzard* days – for the benefit of a newly-joined Chaplain, R.N.V.R.

If insight into the life of the Division in pre-1914 days is provided firstly by such correspondence as remains, a second

THE ILLUSTRATED LONDON NEWS, APRIL 10, 1909.—523

THE TERRITORIAL'S SEAMEN BROTHERS: THE ROYAL NAVAL VOLUNTEERS.

PHOTOGRAPHS BY CLARKE AND HYDE.

1. A GUN'S CREW WORKING THE 4·7 AT THE NAVAL VOLUNTEERS' DRILL HALL AT LAMBETH.
2. HOISTING MASTS DURING BOAT-DRILL.
3. GUN DRILL IN THE PORT BATTERY.
4. SIX-INCH GUN DRILL.
5. SIX-INCH GUN DRILL.
6. FIELD-GUN PRACTICE.

The Royal Naval Volunteers have their floating headquarters on board the old gun-boat "Buzzard," which is moored off Blackfriars. They have also a large drill-hall across the river, at Commercial Road, Lambeth. The instruction given the men is very varied, and includes knots and splices, bends and hitches, wire-splicing, helm and compass, flag and cone, lead and line, boat pulling and sailing, sailmaker's work, etc., in seamanship; rifle and field exercise, heavy gun drill, light quick-firing gun drill, machine gun, gun-loading, Morris-tube practice, and rifle-shooting at Runemede, in gunnery. Instruction is also given in signalling, telegraphy, wireless telegraphy, and in life-saving.

source ought to be personal reminiscences, but the Division is unlucky in that its earliest members left few memoirs (such as Graham's autobiographical *My Ditty Box*) which might assist now. It is a great pity, for London Division was closest to authority and would have had a fine tale to tell.

Survivors from the *Buzzard* days are now very hard to trace. One is Lieut. T. E. Cresswell, who joined London Division on the lower deck in 1910 and who is now 91. He did three fortnights of sea training before 1914, starting with the battleship *Bellerophon* in 1911 during the 'Agadir Crisis', when the fleet went to war routine. He remembers sadly that C.P.O. Hartley fell out of the train on the way home and was killed. 'In those days', Cresswell has recalled, 'not many people had more than a fortnight's leave from work, and it was much to give up. Yet one unlucky party was sent for its fortnight to the battleship *Dominion*, which was in dry-dock at Chatham. Everybody thought that was scandalous. In May 1912 I joined the cruiser *Achilles*, which was sent to St. Kilda. The people there were a terribly interbred stock, and I was appalled to see their condition. Only officers and petty officers were allowed to land'. Cresswell's last pre-war sea training was done in the battleship *Conqueror*, which entertained French sailors at Portsmouth, in support of the then all-important Entente Cordiale. The only sea-time he did on that occasion was the voyage from Portland, where he joined the ship, which was one of the later class of dreadnoughts with 13.5-inch guns in the centre line.

Air Vice-Marshal Sir John Cordingley has left a few papers which are in the Division's archives. He joined the Division in 1905 as a bugler aged fifteen and rose to command the Band, but he resigned as a C.P.O. in 1913, on his marriage, rejoining as a C.P.O. in July 1914 to serve for three years in the Royal Naval Division, in which he obtained a Commission as a Paymaster. He transferred to the R.N.A.S. in 1917 and then to the R.A.F. Cordingley's efforts with the Band were a great success, and the Band was a vital inspiration on ceremonial occasions, as may be deduced from the large number of photographs which feature them.

It is not generally known that on 1 June 1912 the flag of the Commander-in-Chief Mediterranean, Admiral Sir Berkeley Milne, who had taken up the appointment that day, was hoisted in London Division's drill ship, the *President* (formerly *Buzzard*) as she lay off the Victoria Embankment, as the *Times* reported. Two days later, at Portsmouth, it was hoisted in the armoured cruiser *Good Hope*.

The Division's Display of Drills which was held at the drill hall at Commercial Road, Lambeth, on 14 December 1912 was an important occasion, for the first Lord, Mr. Churchill, and Mrs. Churchill were present, he to make the speech and she to give away the prizes. He was in flamboyant mood, and what he said was picturesquely expressed. He said that much of the work that had been done in naval organisation during the past year had been in the direction of securing the swiftest possible mobilisation of the largest number of ships, and that in that operation the R.N.V.R. would not be forgotten. No one could fail to be struck with the quality, intelligence and efficiency of the men of London Division, and though training, intimate knowledge of arms and the long habit of the sea were necessary none of these sufficed without the earnest loyal heart and a resolute and active intelligence. The Volunteers of the Navy, he said, who were intimately associated with it in time of peace, and welded indistinguishably with it should war come, were doing their part, not only to make the country safe and secure, but were also urging forward those great causes to which he had referred.

It was all good speech-day stuff, and it went down well, for there was the First Lord himself acknowledging the worth of the Volunteers. Perhaps his most important statement that day, and the most prophetic, was that 'in the R.N.V.R. the Navy had a very valuable force which it was the duty of the Admiralty to use, if need be, in the very front line'. Many who heard him say that must have recalled his words with bitterness less than two years later when they were sent to do a soldiers' job in the defence of Antwerp as members of the Royal Naval Division. At the time they were spoken, however, they could only take them figuratively, as referring to future sea battles.

The Guard of Honour for the First Lord, and the Small Arms Display that followed, were provided by C.P.O. Cordingley, who said in his appeal for fifty P.O.'s and men to take part 'the drill must be perfect – superior to that performed by the Guards'. So successful were Cordingley's efforts he was asked to give a Field-Gun Display at the H.Q. of both the Queen's Westminsters and the London Scottish.

Admittedly, in the years before 1914, there was no real guidance to indicate how the R.N.V.R. was to be employed when war came, or precisely what was expected of it, beyond a general assurance which was given to London Division by the First Sea Lord, Admiral Prince Louis of Battenberg, on 6 December 1913, when he stated at another Prize-Giving: 'At the Admiralty we have quite recently come to a decision . . . on mobilisation to send to every commissioned ship in the Fleet a detachment of the R.N.V.R.' As Commodore Earl Howe wrote many years later concerning this period, it was intimated that, when the time came, there would be a place waiting in the Service for every efficient member of the R.N.V.R.

The origin of the Royal Naval Division is to be found, as Jerrold said, in plans prepared 'some years' before the outbreak of war, at the instance of the Committee of Imperial Defence, for the formation of a force of Royal Marines to seize, fortify or

Churchill's visit.

protect any temporary naval bases necessary to the use of the Fleet or the provisioning of any army – but the idea of the addition of naval reservists to this force cannot be traced back so easily. In view of the statement of Prince Louis it would appear that it arose after 6 December 1913 at the earliest and before the outbreak of war at the latest, for in Churchill's Introduction (written in April 1923) to Jerrold's *The Royal Naval Division* he confirms its pre-war origin in his statement: 'It was perceived that on mobilisation there would be at least twenty or thirty thousand men belonging to the Reserves of the Royal Navy for whom there would not be room on any ship of war which went to sea'. Perhaps the simplest comment on the subject is that made by Earl Howe in the April 1947 issue of *The Navy*: 'When the day came the Admiralty were found to be in two minds as to how the R.N.V.R. should be employed'.

Two minds or not on 4 August 1914, there was only one on 19 August, for on that day, when Commander Guinness, commanding London Division, and Commander Viscount Curzon (later Earl Howe), commanding Sussex Division, were summoned to the Admiralty for a meeting with the First Lord and other senior officers, Guinness was so shocked and outraged by what he was told that he wrote a first-hand account of it, perhaps for the perusal of the Earl of Selborne, the First Lord who in 1903 had appointed him to command London Division. Guinness wrote:

> Winston Churchill outlined a scheme which he expressed his intention of carrying out. He professed to have discovered, since the mobilisation, the astonishing fact that a

> large proportion of the Naval Reserves could never be required for service with the fleet . . . consequently he intended immediately to fit them for service as a land force. . . . These men were to be hurried into camp in four days, ex-naval officers were to train them as infantry, and later, military officers would be available from the War Office for this purpose . . . . As a force, the Naval Volunteer Reserve was to be destroyed . . . scattered. . . . Their own commanders were to be relieved by ex-naval officers (with the exception of one, a relative of the 1st Lord).
>
> This policy produced consternation among all concerned. Nobody could see that it possessed any advantages. The R.N.V.R. commanders pleaded for the units which they had with infinite pains built up and imbued with a fine esprit de corps. Might they not in any case serve as battalions in the new Admiralty Army? No– 'esprit de corps,' says Mr Churchill, 'cannot be considered in time of war.' A plea is put for the spirit of volunteering which must inevitably be destroyed after such treatment. 'Other things more important are destroyed in times like these', is the answer. Sea Lords of the Admiralty . . . are indignant and assert that there is no justification for the assumption during the first fortnight of what may be a protracted struggle that the Fleet will not need all its Reserves . . . Mr. Churchill waxes eloquent in favour of his scheme, 'It will be a fine force, we will have artillery, and why not cavalry too?' Ye Gods! The Horse Marines at last.
>
> No one approves, no one can find any reason to support the scheme, but alas no one apparently is sufficiently sure of his position to protest. Those in the Force know nothing of what is going on behind the scenes and ask themselves why they have been made catspaws. What was all this praise of their work, of their endeavour to make themselves efficient bluejackets, their smartness and their devotion to the ideal of taking a place, however humble, in the great warships of the first fighting line?[5]

Guinness's astonishing and frank criticism was written in high dudgeon, which is understandable, for he had seen a whole decade of most dedicated perseverance with his London Division betrayed at a stroke. Although Churchill had claimed he had no alternative but to use the Naval Volunteers as a land force, the Earl of Selborne later denied this in a letter to Guinness, saying that in the Special Reserves and Territorials there were available plenty of 'troops who were better equipped, better trained and more effective for the purpose'. In a further reference to the interview he had with Churchill, Guinness wrote: 'He sent for me, not to consult me, but because he wanted me to say what a fine scheme it was and I am afraid I said I thought it rotten.' Writing thirty-eight years later, Earl Howe, who had been present at the interview, said that by the decision taken in August 1914 the R.N.V.R. had been made the only conscripts in all the British forces. In 1923 (in the Jerrold Introduction) Churchill admitted that losses at sea and expansion of the Fleet by early 1915 would have provided ample opportunity for the use in the Navy of all in the R.N.V.R.; yet he had said: 'a large proportion . . . could never be required'.

It is clear from what Guinness wrote that to employ these men in land fighting with the Royal Naval Division was the result of a hasty and capricious judgment by Churchill, made in his most bellicose mood, as Asquith said, which overrode that of the Sea Lords. Churchill had been a valiant cavalry officer, and when he sent the Royal Naval Division to defend Antwerp in October 1914 he offered to resign from the Admiralty and take personal command there, but Mr. Asquith would not hear of it. In view of that proposal it is not inconceivable that he dreamed of commanding his Royal Naval Division in other and more famous theatres of war. After all, Kitchener had said he was willing to make him a Major-General, and a large number of those who took up 'temporary' R.N.V.R. commissions were privileged and/or distinguished young men – such as Arthur Asquith, Rupert Brooke, Denis Browne, Bernard Freyberg, Cleg Kelly, Patrick Shaw-Stewart and Charles Lister. The 'relative of the 1st Lord' to whom Guinness referred in his account of the Churchill interview was doubtless Lt. Col. George Cornwallis-West, R.M., Commandant of the Anson Battalion of the Royal Naval Division, a former Scots Guards officer who married Churchill's widowed mother.

What must surely be seen as the final item of the pre-war history of the R.N.V.R. is incredible – even those members who happened to be undergoing their sea-training in ships at the end of July 1914 were put ashore, and, together with the majority of the mobilised R.N.V.R. men, 'pitchforked into the Royal Naval Division'.

CHAPTER THREE

Called to Arms

Early in August 1914 London Division consisted of over a thousand officers and men – 1332 to be precise. It had mobilised 804, welcomed back 216 who had rejoined, and recruited 312 newcomers. In the course of time, however, 258 from the overall total were to serve in other units. Two or three weeks later, when the Division had drafted away its members, it ceased to function, not merely for the duration of hostilities but also for two or three years beyond their cessation, in common with the five other Divisions (Clyde, Bristol, Mersey, Sussex and Tyneside). Thus from August 1914 to April 1921 London Division, like happiness, has no history, although those who belonged to it remained on its books for *Navy List* purposes, with its forty officers (including Honoraries) corporately included, from Commander Guinness of 1903 to Midshipman Campion who joined as late as 13 July 1914. Spiritually, if not physically and

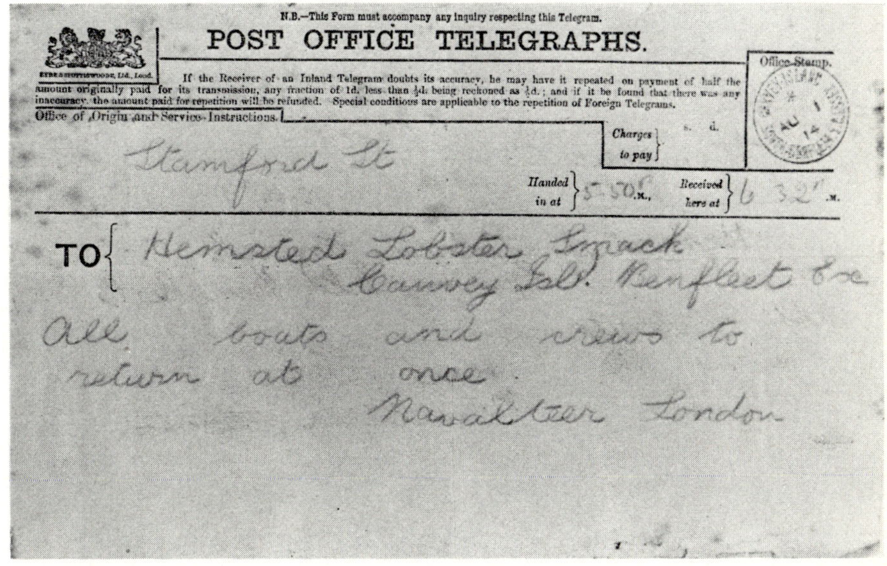

N.B.—This Form must accompany any inquiry respecting this Telegram.

POST OFFICE TELEGRAPHS.

If the Receiver of an Inland Telegram doubts its accuracy, he may have it repeated on payment of half the amount originally paid for its transmission, any fraction of 1d. less than ½d. being reckoned as ½d.; and if it be found that there was any inaccuracy, the amount paid for repetition will be refunded. Special conditions are applicable to the repetition of Foreign Telegrams.

Office Stamp.

Office of Origin and Service Instructions. Stamford St

Charges to pay s. d.

Handed in at 5 50 .M., Received here at 6 32 .M.

TO { Hemsted Lobster Smack
Canvey Isl. Benfleet Esx

All boats and crews to return at once.
Naval Ver London

The original telegram received by Yeoman of Signals Hemsted whilst instructing during a sailing weekend over the August Bank Holiday in 1914. The Lobster Smack Inn was regularly used as a base for these weekends.

AUGUST 8, 1914]

THE BRITISH FORCES : *Scenes in London and the South on Sunday and Monday, August 2 and 3*

The Mobilisation of the Royal Volunteer Naval Reserve on Sunday

On Sunday last the naval authorities ordered the mobilisation of the Royal Naval Reserve and the Roya Fleet Reserve, and at the present moment every reserve man is at his post of duty. The Volunteer Nava Reserve were mobilised at their headquarters, where they received their detailed instructions. Later they were inspected by the Hon. Rupert Guinness. The above picture shows Mr. Guinness and the reserve men

Waiting for the call.

geographically, they were still a band of brothers, many of whom would meet and serve together again when peace had been restored. In the meantime London's thousand or so of the R.N.V.R. were helping to make new Royal Navy history in ships and establishments spread far and wide, and particularly in the Royal Naval Division, in which, though for two years they were under Admiralty control, they fought as soldiers.

The telegrams went out and London Division were mustered with smoothness and precision. There were neither beds nor facilities for feeding them in the Commercial Road drill hall, and, without instructions from the Admiralty, Guinness improvised. He bought supplies of food, straw for palliasses, other equipment, housed the men in nearby L.C.C. schools, and put his wife, Lady Gwendolen, in charge of the catering. As the volunteers' pay was not immediately forthcoming he advanced large sums of money to them. So, supported by their Commanding Officer, the volunteers all waited for the expected order which would draft them to the Fleet. It never came.

The flavour of those exciting days has been well remembered by Able Seaman H. J. Dyer, who has written: 'When war was declared I was at a Scout Camp at Margate. On the sea front the news was flashed out on an illuminated strip and as dusk was falling we watched an array of ships of the Royal Navy steaming up the English Channel on their way to battle stations, and the local bands playing "When our Ships Put Out to Sea". The impact of the declaration of war did not register at that time to myself, but in the morning an urgent telegram from my mother informed me that I was at war in the Royal Navy. A Scout guard of honour carried my kit-bag on poles to the station where I was given a bugle send-off. In London I changed into

Navy gear and off to the drill-hall. Our beds had not arrived, so we had twenty-four hours' leave. When the camp beds arrived we waited events. One morning a few select members left for their ship. I remember joining a party that manhandled a low truck over the bridge to the Navy Depot at Deptford to collect Navy rations and supplies. We were in the Navy. Then the bombshell! Churchill had turned us into soldiers. We were now a Naval Brigade.'

When hostilities began, London Division had thirty-seven officers (excluding Honoraries), of whom twenty-four were Executive and ten were Staff Surgeons or Surgeons. Not all appear to have received appointments immediately, but, of those who did, by December 1914 only one Executive officer had gone to sea, while sixteen had been appointed to the Royal Naval Division. Of the Surgeons, who were something of a special case, seven had gone to sea, including five to hospital ships. Thus out of the thirty-two London Division officers who had been given appointments by December 1914 only three were in warships – approximately 10%. Yet the twenty-four Executive officers had between them completed three Courses of Strategy and thirty Short Courses in Gunnery, Torpedo or Signalling, passed two Board of Trade Examinations in Navigation, and in sixteen instances had served in the Fleet for short cruises of not less than one month, all in their off-duty time. The only officer of the Executive branch who had been given a sea appointment was Lieut. G. Frecheville, and he had gone to the battleship *Iron Duke*, flagship of the Grand Fleet.

The Division's lower deck had had a somewhat similar experience. A small percentage, particularly Signalmen, went to ships, but the majority were drafted willy-nilly into the Royal Naval Division. A censored postcard dated 7 July 1915 was sent from the battleship *Albemarle* to Yeoman of Signals J.R. Hemsted in the *Iron Duke* by his former colleagues in the *President*. It reads: 'The boys all send kind regards to you. Young, Hand, Connell and myself are in the *Albemarle*, Butler and Jenkins are in the *Russell* and the rest in the *Diamond*, *Topaze*, *Venerable* and *Exmouth*. Life is strange but we are dropping into it slowly.' These men were among the lucky Signalmen, and they wanted to tell their old Yeoman what had happened to them. Hemsted had been a leading figure in the pre-war days in the *President* and he had a great career ahead of him. He was commissioned in due course and ended the war as a lieutenant, going on to command London Division from 1933 to 1939 as a Captain.

The 'force of Royal Marines' planned by the Committee of Imperial Defence before the war 'to seize, fortify or protect' any temporary naval bases had already begun to form in four battalions by 16 August 1914. On that date the First Lord, Mr. Churchill, took the decision to add to it two naval brigades of 'surplus naval reservists of different classes', thus establishing the Royal Naval Division. Half the personnel of the Marine Brigade were active service men. The design was that each of the eight battalions of the two naval brigades should have 375 Royal Naval Volunteer Reserve, 315 Royal Fleet Reserve and 190 Royal Naval Reserve men, giving a total of 3000 R.N.V.R. men in the two brigades.

Thus soon after mid-August 1914 most of the seamen of the new Royal Naval Division, who were under canvas near Walmer and Betteshanger in Kent, found themselves in two brigades, the First Naval Brigade having its Drake, Hawke, Benbow and Collingwood Battalions, and the Second Naval Brigade its Nelson, Howe, Hood and Anson Battalions. Whereas Churchill's order had said that three of the eight battalions were to be commanded by Commanders, R.N.V.R., only Commander Viscount Curzon (later Earl Howe) had been so honoured by the end of December 1914. R.N.V.R. participation in each battalion was increased to 424 men. A London Division member of the Drake Battalion has written: 'We were a strange amalgam, joined by Royal Naval Reserve and Royal Fleet Reserve and an invasion from the Scottish Isles of a Gaelic-speaking community, where, if one shouted "MacDonald", the whole clan erupted.'

During August and September the officers and men of the Royal Naval Division awaited a call to action as the German Army descended on Belgium, and the British Expeditionary Force fought in the Retreat from Mons and then in the battles of the Marne and the Aisne. Following the Aisne, attention was turned to the Channel Ports which because of their importance to the British supply lines were a prime element in British naval strategy. It seemed that the very sort of situation for which the Royal Naval Division had been raised was looming up across the Channel.

On 2 October the Germans broke through near Antwerp, driving back the Belgian Army. If Antwerp was not defended the Allied line was in danger of being outflanked, and this was unacceptable for naval reasons alone. That evening Churchill visited Antwerp, and two days later the Marine Brigade entered the city and moved into trenches. Churchill had also undertaken to send out the two Naval Brigades, which were untrained and still under canvas. So they marched to Dover, loaded their own stores and crossed over, starting, as one man told, at 3 a.m. He was in the *Mount Temple*, the C.P.R. ship which had been in the vicinity when the *Titanic* sank in 1912, and she arrived at Dunkirk at 8 a.m. Sleep had been possible for a few, but for many the overcrowding was so bad that the men were compelled to stand almost the whole time they were on board. They had had only a makeshift meal and had to unload their ship on arrival and then to load up the waiting trains. They had been

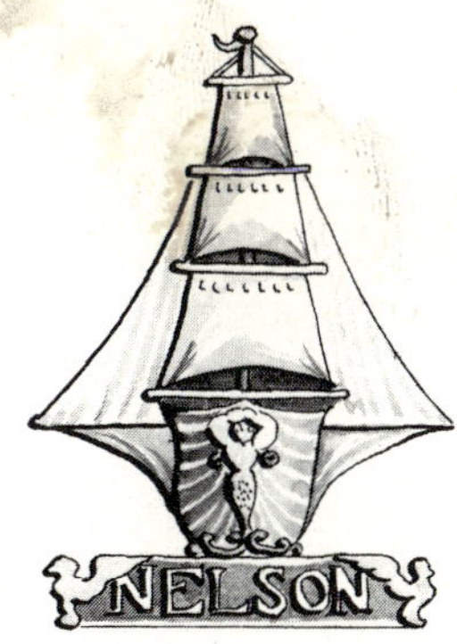

issued with 120 rounds of ammunition each, which had to be carried in pockets as few had bandoliers or haversacks. By the time they entrained the men were exhausted. Rupert Brooke, the poet, who was serving in the Anson Battalion, described the experience as a 'very tragic and amusing affair'.

Years later, at a Royal Naval Division reunion, Air Vice-Marshal Sir John Cordingley recalled this Channel crossing, and told how, on finding no nominal lists, he took steps to try and bring order out of a chaotic situation. He believed this initiative earned him a reputation for administration which later led to his selection to assist in the setting up of the Royal Naval Air Service and ultimately to his distinguished career in the R.A.F.

The Naval Brigades got a fine reception in Antwerp, and as they marched they were given food and drink by happy Belgians. Able Seaman A. W. St. C. Tisdall of the Anson Battalion, of whom much more will be recounted later, wrote from Antwerp on 7 October saying: 'The singing of shells is indescribably weird when you first hear it. All our kits lost by fire. The burning city of Antwerp is a terrible but magnificent sight against the blackness of the night, and lights up the whole country round. The sight of the poor women and children driven from their homes makes one's blood boil. It's horrid to feel so useless.'

In the situation the British faced at Antwerp the Naval Brigades were a slender asset militarily, being without orders and equipment. They did not know until they arrived at Antwerp that they were to come under the command of General Paris who commanded the Marine Brigade. Deteriorating morale in the Belgian Army soon compelled General Paris to

Royal Naval Division building trenches during the defence of Antwerp. October 1914.

withdraw to the city's inner defences. There the Naval Brigades, woefully short of sleep and food, occupied trenches, where a few, including the officer commanding the Collingwoods, were killed, and some were wounded, in a desultory bombardment. Eventually, continual defections by Belgian fortress troops led the General by 6 p.m. on 8 October to order the Royal Naval Division to retire to Zwyndrecht, fourteen miles west of Antwerp, the 1st Brigade by one route and the 2nd and Marine brigades by another. The Portsmouth Battalion of the Marines was to act as rearguard. All were to leave *on receipt of the order*.

The order intended for Commodore Henderson, who commanded the 1st Brigade, was wrongly delivered to the C.O. of its Drake Battalion, who was told that the Brigade had already been informed. Henderson only learnt of the order about an hour later from a staff officer who was returning from taking the order to the Marine Brigade. Unfortunately the need for instant departure was not mentioned, so Henderson and his Hawke, Benbow and Collingwood Battalions left for Zwyndrecht after 9.30 p.m., along roads choked with civilian refugees. The 2nd Brigade, the Marine Brigade and the Drake Battalion (of the 1st Brigade) arrived there six hours after their much earlier start, and marched on to safety, but the 1st Brigade was held up by a destroyed bridge, paused for two precious hours in Zwyndrecht seeking information and lost all cohesion.

Finally, acting on a report which later proved to be false, Commodore Henderson, in order to save needless casualties among his men and inevitable surrender, crossed the Dutch frontier with his brigade staff and his Hawke, Benbow and Collingwood Battalions, all of whom were interned for the duration of the war. Some naval stragglers who had managed to entrain with the rearguard Marine Battalion west of Zwyndrecht escaped when their train was attacked, and an R.N.V.R. officer and forty men of the Benbow Battalion also escaped by making their way along the Dutch frontier.

The losses suffered by the Royal Naval Division were severe. Seven officers and fifty-three men had been killed; three officers and 135 men had been wounded; thirty-seven officers and 1,442 men had been interned, and five officers and 931 men had been captured by the enemy. There was a public outcry when the losses became known.

The Royal Naval Division internees were sent to Groningen in northern Holland, where they were placed in the military barracks and guarded by Dutch sentries. The officers, who were on parole, went to the barracks daily to look after their men and to try to cheer them up. Initially, the Dutch authorities only allowed the men out for route marches round the canals, on the basis of one battalion a day, but later the Dutch Minister of War decided to allow fifty men to go out on leave each afternoon, with one Dutch soldier to every six sailors.

The camp at Groningen, Holland, known as 'HMS Timbertown' where members of the Royal Naval Division, including several hundred from London, were interned after the fall of Antwerp.

Early in 1915 the internees were moved into wooden hutments built on a recreation-ground opposite the barracks, and in time they all agreed it was a change for the better. The whole settlement soon received the name of Timbertown. Bound volumes of its *The Camp Magazine* which was skilfully produced throughout the period of internment are preserved in the archives of London Division, to which a fair number of the men belonged.

Remarkably, three of the internees from London Division, H. J. Selkirk, R. H. Guppy and J. D. Ewings, lost their lives through enemy action at sea. They were pronounced missing while returning to Holland and internment from compassionate leave in England. Their ship, the S.S. *Copenhagen*, was torpedoed by a U-boat near the N. Hinder light-vessel on 5 March 1917. *The Camp Magazine* of May 1917 carried a black-bordered commemoration which spoke of 'the dangers of a perilous crossing'.

In welcoming back the Royal Naval Division on their return to England Mr. Churchill said 'It is too early yet to judge what effect the delaying, even for five or six days, of at least 60,000 Germans before Antwerp may have had upon the fortunes of the general battle to the southward. It was certainly powerful and helpful'. He also said that the loss of a portion of the 1st Brigade, through a mistake, in no way reflected upon the quality or the character of the Division, whose instruction and training had been 'interrupted, for a time'. Further plans for its use on other fields were obviously being made, in accordance with the First Lord's minute of 16 August governing the formation of the Royal Naval Division which had ended with these ominous words:

> All the men, whether sailors or Marines, while training in the three brigades will be available if required for service afloat; but *in the meanwhile* they will be left to be organised for land service.[1]

That 'meanwhile' was to last for the duration of hostilities (the italics are supplied by the author). As it transpired, the Division was yet to win immortal fame.

In mid-September 1914 the huge Crystal Palace at Sydenham had been put to use as a training Depot for the Royal Naval Division, and Commodore Sir Richard Williams-Bulkeley, Bt., R.N.R., who had formerly been in command of Mersey Division, R.N.V.R., was its Commanding Officer. Certain officers of London Division held important appointments there, including Roberts-Wray as Brigade Major, Wildy as Drafting Commander and Greenwood and Nisbet as Paymasters. J. H. Mather, who became an Acting Lieutenant on 12 October 1914, had been a member of the Ship's Party in the *Terra Nova* on Scott's Antarctic Expedition of 1910, in which, as a Petty Officer, R.N.V.R. (the only R.N.V.R. man on board) he was taxidermist. At the Crystal Palace he was put in charge of the Signal School. He later became a Commander.

From the outbreak of war there had, of course, been large numbers of volunteers coming forward for service in the Navy and Army, and the pick of these had found their way, in some cases clearly through interest, into the Royal Navy, where they quickly obtained commissions as Temporary Officers in the R.N.V.R., without having had a previous connection with any of its local Divisions. In addition, many who had been serving as ratings in those Divisions were awarded temporary commissions. A good instance of this was Tisdall. He received his commission in the Anson Battalion of the Royal Naval Division (dated 1 October, before he went to Antwerp) immediately on his return on 11 October. By January 1915 the number of officers holding such temporary commissions was well over 400, and, of these,

204 were serving with the Royal Naval Division and 172 were appointed to H.M.S. *Pembroke* (i.e. Chatham) for Special Service with the Royal Naval Air Service.

Despite the formal cessation of the work of London Division on the outbreak of war it appears that recruitment of ratings for the R.N.V.R. took place for some weeks at least at the drill hall at Commercial Road, Lambeth, and that those who so entered were acknowledged to be of the Division, their Service Certificates being signed by the Commanding Officer, Commander Guinness. A representative of this group is Arthur Watts, who became Recruit No. 97 on 3 September 1914 with the intention of serving only for the war. He joined with his brother, John, and they spent about fourteen days at Commercial Road before being put in the first party to be sent to the Crystal Palace, still wearing the plain clothes of civilian life as no uniforms had yet been issued. There, all had to sleep on bare floorboards for a week or two in what was called the 'Canada Building' until hammocks were supplied. Arthur and John Watts were made Leading Seamen on 18 September and were included in the first Officer's Class which the Crystal Palace Depot produced (and which was in fact the first C.W. Class in the R.N.V.R.). Both were commissioned on 29 December, put into lodgings outside the Crystal Palace for a week or two, and then appointed to the Collingwood Battalion of the Royal Naval Division at Blandford Camp in Dorset.

On 10 May they sailed for the Gallipoli Campaign in the troopship *Ivernia*, and about eighteen days later went ashore at night at 'V' Beach, alongside the *River Clyde* which had played such an important part in the desperate landings on 25 April. A few days later, in the Collingwoods' attack on 4 June in which the Battalion was destroyed, Arthur Watts was severely wounded. His brother John and K. W. J. Oldridge, also of London Division, were the only two officers who survived unscathed in that disastrous attack, in which sixteen officers were killed, eight were wounded and there were well over 500 casualties among the men. Lieutenant and Quartermaster Stanley Geary, R.M., a former C.P.O. of London Division, was wounded in the early morning of 6 June in a Turkish artillery bombardment. Arthur Watts was sent back to England via Alexandria in a hospital ship, and after he had been fitted with an artificial leg at Roehampton presented himself to the authorities at the Crystal Palace, where he demanded a new appointment and was eventually seen by the Commodore. His persistence was finally rewarded when he became one of the first R.N.V.R. Intelligence Officers at the Admiralty.

In 1982, at the age of eighty-eight, Arthur Watts presented his sword to London Division, where it is reserved for the use of the Commanding Officer. The sword was originally given to him, to mark his receiving his commission, by his father, who had served in the old Royal Naval Artillery Volunteers.

A further representative of the group of temporary R.N.V.R. ratings who were recruited through the Division was C. G. Bonner, whose name Watts pencilled into a small pocket book (which still exists) as being one of his first messmates in September 1914. Bonner's name is crossed out, however, because, after only a few weeks, he was drafted away when it was found he was a Master Mariner. On 8 August 1917 Bonner was one of the two who were awarded the Victoria Cross in the 'Q' ship *Dunraven*, commanded by Captain Gordon Campbell, after an action with a German submarine. At that time Bonner was a Lieutenant, R.N.R. Bonner's V.C. is noteworthy in that it was announced in the Court Circular issued from York Cottage, Sandringham, on Sunday, 8 October 1917, before the award was announced in the *London Gazette*, and it was on this day that he received the Cross, perhaps the only occasion when a man received his V.C. from the Sovereign's hand on a Sunday[2].

As soon as possible after the return of the Royal Naval Division from Antwerp the Staff at the Crystal Palace had the urgent task of rebuilding the lost or depleted battalions, a process which meant the percentage of participation by London Division, R.N.V.R., became much smaller. To take the Collingwood Battalion as an example, of the original 700 or so of its R.N.V.R., R.F.R. and R.N.R. members who went to Antwerp only twenty-two returned to England, the remainder being either interned in Holland or prisoners in Germany. The twenty-two survivors became merged in an entirely new Collingwood Battalion, which had a huge preponderance of Durham and Northumberland miners, all of whom had enlisted voluntarily. A commemorative brochure about the Collingwoods which was produced by Stanley Geary gives the home addresses of all members, and London, and even southern, addresses are only rarely found.[3] One London Division man who continued in the new Collingwoods, and who survived yet again in the slaughter of 4 June 1915 at Gallipoli, was Leading Seaman A. R. Blore, who was awarded the Conspicuous Gallantry Medal in that action. Later, when he was serving in France in the Anson Battalion of the Royal Naval Division, Blore was given a Bar to that medal, thereby becoming the only man in the Royal Navy to be so honoured.

The plan to concentrate the Royal Naval Division at Blandford Camp in Dorset began to take effect in November 1914. In December active service uniform was issued to all ranks, though naval caps and tallies and naval cap badges and distinction lace for officers were retained despite the khaki. The change was a great aid to discipline, even if it led eventually in France to the rumpus over 'Codner's Beard', about which A. P. Herbert, then a Lieutenant in the Hawke Battalion, submitted a poem to *Punch* (*The Ballad of Codson's Beard*, 1917 – see

A. E. Watts as a rating.

Sub. Lieuts. A. E. and J. C. Watts in the uniform of the Royal Naval Division at Blandford Camp, Dorset prior to sailing for Gallipoli. (Note the Sub. Lieutenant's ring round the sleeve and naval cap badge on army style uniform.)

Appendix).[4] The Navy regarded its right to 'grow' as well-nigh sacred, but increasing militarisation in 1916 led to a showdown. 'If we had lost Codner's beard, for it became soon far more ours than his, we lost everything,' wrote Jerrold in his *The Royal Naval Division*.

By the end of January 1915 the Royal Naval Division, less the Hawke, Benbow and Collingwood Battalions, which were still training, was at Blandford awaiting orders. About a month later it was inspected by the First Lord, Mr. Churchill, and then, two hours afterwards, by King George V. Arthur Watts had the responsibility of getting the special flagstaff erected which would fly the Royal Standard on that occasion. It arrived without halliards, but when none could be found on a cold and snowy day it was erected without these necessities after a volunteer had offered to shin up the flagstaff and put them in place when they should appear. On the great day all was well, but in the meantime Watts had had to explain the strange procedure to his battalion commander, Commander Spearman, R.N. Halliards or no, such high-powered inspections almost on the eve of departure overseas ensured that the legendary reputation of the Royal Naval Division was under way.

On 28 February the Division embarked at Avonmouth in the *Franconia*, the *Braemar Castle*, the *Gloucester Castle*, the *Grantully Castle*, the *Minnetonka* and the *Astrian*, bound for Gallipoli. Lady Violet Bonham-Carter, the daughter of the then Prime Minister, Mr. Asquith, has told in her *Winston Churchill As I Knew Him* how she saw off her many friends, including her brother, in the *Grantully Castle*: 'The decks were densely crowded with young, splendid figures, happy, resolute and confident and the thought of the Athenian expedition against Syracuse flashed irresistibly through my mind.'

The five landings on the Gallipoli Peninsula took place on 25

The Royal Naval Division landing at Gallipoli.

April, and in four of these the Anson Battalion of the Royal Naval Division and the Plymouth Battalion R.M.L.I. were concerned, respectively at 'V' Beach ('A' Coy. and 1 platoon of 'D' Coy.), 'W' Beach ('B' Coy. and the Anson Battalion H.Q.), 'X' Beach ('D' Coy., less 1 platoon, and 'C' Coy.) and at 'Y' Beach (Plymouth Battalion R.M.L.I.). Three hundred ratings with sea experience, who were drawn from the Hood and Howe Battalions of the R.N.D., carried out beach duties in connection with the landings, under Sub-Lieut. J. B. Dodge. These ratings manned the tugs and lighters which played their part in getting the landing parties ashore from open boats. A platoon of the Ansons manned the boats taking the Dublin Fusiliers inshore in six tows in line abreast on the left of the *River Clyde*, in which other Dublins, the Munster Fusiliers, half the Hampshire Regiment and Sub-Lieut. Tisdall's platoon of the Ansons were carried. When the *River Clyde* grounded, a delay occurred in fixing up the 'bridge' of lighters. The Dublins, landing from their open boats, were almost annihilated, but about 150 reached the shore alive, to find what cover was to be had or to enter the ruins of Sedd-el-bahr fort.

No tide of men emerged from the *River Clyde* because of the hail of fire, and only ones and twos succeeded; for every one who stepped ashore from the floating bridge of lighters six or seven fell in the fifty yards which separated the ship from the shore. The wounded fell from the lighters and lay in the water, beyond help. Commander Unwin, R.N., and his officers from the *River Clyde* worked hard for hours on the lighters and in the water, making fast, repairing damage and then rescuing wounded. Unwin was assisted also by Tisdall's Anson Battalion detachment. Tisdall jumped into the water and pushing a boat in front of him went to assist the wounded. Obliged to obtain help, he took with him on two trips Leading Seaman Malia and on others Chief Petty Officer Perring, Leading Seaman Curtiss and Leading Seaman Parkinson. Tisdall made four or five trips to the shore and rescued several wounded men under heavy and accurate fire.

A letter in the *Times* of 6 December 1915 first brought to public notice Tisdall's consummate bravery, the writer stating: 'I have never seen more daring and gallant things performed by any man, naval or military, than those performed by the man I now know to have been Sub-Lieut. A. W. St. C. Tisdall, Anson Battalion, R.N.D. at the landing from the *River Clyde* . . . a boat containing an officer (unknown to all) and three bluejackets, one of them a petty officer, was very prominent . . . Darkness came on, and that officer was nowhere to be found. All the petty officer and bluejackets could say was, "He's one of those Naval Division gents." After much trying to find out who the officer was, the writer received a note from the 'V' Beach trenches stating the only R.N.D. officer on the *River Clyde* had been Tisdall.'

A photograph taken by a Lieutenant of the Hampshires from the *River Clyde* captured the moment when the boat Tisdall used was seen at the shoreline doing its task of mercy, and in this photograph about 200 men can be seen just beyond the boat sheltering under a bank having reached the shore. This photograph is reproduced in this study from a small book which was written about Tisdall in 1916.[5] On 31 March 1916 Tisdall

Taken at V beach 25 April 1915 from the River Clyde *at the moment when Sub. Lieut. Tisdall was winning his VC.*

was awarded the Victoria Cross for 'most conspicuous bravery and devotion to duty'. The citation went on to say, after describing the deed, 'Owing to the fact that Sub-Lieut. Tisdall and the platoon under his orders were on detached service at the time, and that this officer was killed on May 6, it has only now been possible to obtain complete information as to the individuals who took part in this gallant act.'

King George V later wrote of his regret that Tisdall's death had deprived him of the pride he would have had in personally conferring the honour on Tisdall. A naval surgeon wrote in the *Times* on 8 December 1915 to tell how Tisdall died. He was standing on the parapet of a trench where his men were taking temporary cover, and he was buried close to where he fell. He disdained to take cover himself. The surgeon had met Tisdall two days after his gallant deed, and Tisdall had said 'I was never in such a funk in my life.' Tisdall was a Scholar of Trinity College, Cambridge, and in his last year there won the Chancellor's Gold Medal. He joined London Division on the lower deck on 15 January 1914.

At 'W' Beach the attack was instantly successful. This was 'Lancashire Landing', where the Lancashire Fusiliers with 'B' Company of the Anson Battalion of the R.N.D. approached the shore in four lines of cutters, eight abreast, towed by eight picket boats. When the tows were cast off the boats were rowed ashore. The main body of the Lancashires cut their way through the wire, reorganised under the cliffs and captured the enemy trenches. The Anson platoons and the Hood and Howe detachments took part in this fighting, before being re-formed and re-assigned to their duties as beach parties. Sixty-three out of eighty naval ratings were killed or wounded. This was the morning when the Lancashire Fusiliers won their 'six V.C.s before breakfast', a most incredible feat of arms. Throughout the day the Ansons at 'W' Beach were occupied unloading stores, transporting ammunition and burying the dead, the company which had shared in the first assault having been reinforced by three platoons of 'A' Company diverted from 'V' Beach. The 'C' and 'D' Companies of the Ansons performed similar duties at 'X' Beach.

The main body of the Royal Naval Division in their troopships took part in a feint attack much farther up the Gallipoli Peninsula in the direction of Bulair in the Gulf of Xeros, but the Drake battalion was soon detached from this movement and sent to Cape Helles, where they were in action on 28 April. By 4 May all of the Royal Naval Division were ashore, and within four days they had sustained some heavy losses. The 2nd Brigade under Commodore Backhouse did some fine work. Before the end of May, the awaited Hawke, Benbow and Collingwood Battalions arrived from England, and General Paris had his Division concentrated at last. The 'big push' took

A bayonet charge of the Royal Naval Division at Gallipoli.

place on 4 June at noon, with French troops on the right. When these evacuated positions which they had just gained, the Collingwood Battalion was enfiladed and almost annihilated.

In December the final decision was taken to evacuate the Gallipoli Peninsula, at which time the Royal Naval Division was at Cape Helles. The withdrawal was complete by 3.30 a.m. on 9 January 1916, but there was an R.N.D. presence until 3 a.m. when General Paris and his staff embarked.

In May 1916 the Royal Naval Division went to France, and, within the long period of the battle of the Somme, fighting now under Army and not Admiralty orders, distinguished itself at the battle of the Ancre, and particularly at Beaucourt. Thereafter it saw heavy fighting at Gavrelle, Passchendaele and Welsh Ridge, before participating in the bloody Allied defensive in the spring of 1918, the attack on the Hindenburg Line and the final advance to Victory. In the November 1916 attack in the fog and mist at Beaucourt in the Ancre Valley, the Royal Naval Division lost more than a third of its strength, while advancing further and taking more prisoners than any division had done in one day, as Sir Douglas Haig said. In the course of all the action in France the number of those who had gone into the Division from the pre-war personnel of the *President* had reduced to vestigial proportions. One officer of the R.N.D. who came from London Division, and who remained to the end, was Commander A. W. Buckle, who won a D.S.O. and no less than three bars. Churchill described him, along with six others including Col. Freyberg, V.C., who as Lieut-Cdr. Freyberg had commanded the Hood Battalion from 1915, and Commander Beak, V.C., as 'salamanders born in the furnace'.

The Royal Naval Division, which had been inspected by King George V just before it went overseas to Gallipoli in 1915, held its final parade on Horse Guards Parade on 6 June 1919 (exactly twenty-five years to D-Day) and was inspected by the then Prince of Wales, who in his address said its achievements had been 'worthy of the best traditions both of the Royal Navy and the British Army', but observed that few who were listening to him could have been among those to whom the King had bade farewell. The report of this occasion concludes by commenting 'They were then dispersed to their respective Divisions' – i.e. of the R.N.V.R. – but that was just the theory, for the new existence of those Divisions had not yet been thought out.

A fine memorial obelisk designed by Sir Edwin Lutyens, bearing on one side Rupert Brooke's sonnet beginning 'If I should die, think only this of me', which had been written at Blandford Camp while Brooke was sharing a room with Lieut.-Cdr. Freyberg, was erected on Horse Guards Parade at the south-west corner of the Admiralty and unveiled in 1925. It commemorates no less than 582 officers and 10,295 other ranks of the Royal Naval Division who gave their lives. The obelisk had to be removed when the Admiralty Citadel was built during the Second World War, but it was re-erected later at the Royal Naval College at Greenwich and committed to the keeping of the Royal Navy. At its first unveiling Mr. Churchill said the annals of the Royal Navy contained no brighter page than that which the Division had contributed to them. In his Introduction to the history of the R.N.D. which was written by Douglas Jerrold he said: 'Deriving as they did their nomenclature, their ceremonial, their traditions, their inspiration from the Royal Navy, they in their turn cast back a new lustre on that mighty parent body of which it will ever be proud and for which it must ever be grateful.'

The Royal Naval Division Association, founded in 1920, held its Final Reunion at Greenwich on 31 May 1981, when twenty-three survivors attended, supported by a host of official and private guests, among whom were Churchill's grandson, Mr. Winston Churchill, M.P., the Commanding Officer of London Division and the Turkish Military Attaché, for old time's sake. The Second Sea Lord, the Commandant General of the Royal Marines and the Admiral President of the Royal Naval College were also present, while among the lining party were ten young members of London Division's New Entry Class. Messages were received from Her Majesty the Queen and the Admiralty Board. Each of the surviving members of the Royal Naval Division later accepted an invitation to become an Honorary Member of London Division's 'Old Hands' Association, so the link with London Division will be preserved.

Despite the long and most honourable tale of the Royal Naval Division and the legendary fame which it won for itself on several fronts, the decision to turn into soldiers men whose whole training, 'prolonged if partial', as Churchill wrote, had been modelled on that of the Senior Service has never been justified. It was sheer waste of good material, which proved to be so sadly needed at a later stage of the war, when good use could have been made of so many in the Royal Naval Division who had been killed, captured or interned. That they achieved such great things in spite of their injured feelings about not going into ships on the outbreak of war proves what dedicated men they were. For many, a decade or more of naval training was thrown on the scrap-heap overnight. A corps less loyal 'might not have stopped at deep and bitter cursing', as Earl Howe later judged. When the decision to turn them into soldiers was first announced, resentment in London Division found vent in the ceremonial burying of a marline-spike and a copy of the *Seamanship Manual*, while Commander Guinness hung placards on the Division's guns saying 'Not Wanted In This War'. At the same time as this was happening, the Admiralty, to make up complements of newly-commissioned ships, swept bare the Gunnery, Torpedo, Signal and Navigation Schools of instructional personnel, although later the dislocation thus caused had to be slowly overcome by re-staffing the Schools. Only a few R.N.V.R. officers and men were able to find their way to sea. There obviously was a mental block operating in official thinking about the R.N.V.R., and it may have been a persistence of the kind of thinking which disbanded the R.N. Artillery Volunteers in 1892. It is hard to explain it in any other way. No effort was made to appoint men to long technical courses, nor did the Admiralty show itself eager to open commissioned ranks to those men who showed qualities of leadership.

A Gallipoli medallion.

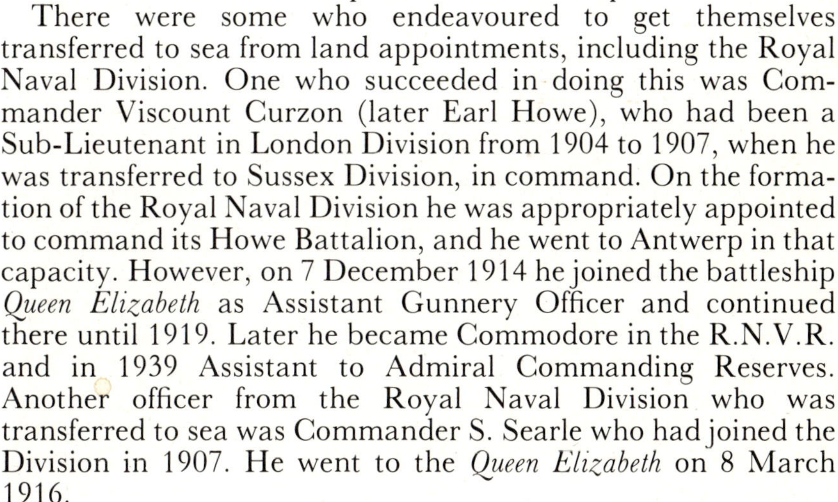

There were some who endeavoured to get themselves transferred to sea from land appointments, including the Royal Naval Division. One who succeeded in doing this was Commander Viscount Curzon (later Earl Howe), who had been a Sub-Lieutenant in London Division from 1904 to 1907, when he was transferred to Sussex Division, in command. On the formation of the Royal Naval Division he was appropriately appointed to command its Howe Battalion, and he went to Antwerp in that capacity. However, on 7 December 1914 he joined the battleship *Queen Elizabeth* as Assistant Gunnery Officer and continued there until 1919. Later he became Commodore in the R.N.V.R. and in 1939 Assistant to Admiral Commanding Reserves. Another officer from the Royal Naval Division who was transferred to sea was Commander S. Searle who had joined the Division in 1907. He went to the *Queen Elizabeth* on 8 March 1916.

CHAPTER FOUR

The Fragile Peace

In March 1918, when British fortunes were at a low ebb during the spring offensive by the Germans, and Victory still seemed far away, the drill-ship *President* came to life again, when the boys of the Marine Society's training ship *Warspite* took over its echoing emptiness when their own floating home, moored at Greenhithe, was destroyed by fire. Large crowds on the Embankment watched the evolutions of the boys, after the manner in which, years before, other crowds had watched the men of the R.N.V.R. To many present, the sight must have augured well for the days to come, when those men would return from the war and continue their service in the peace which they had helped to win.

Three years, however, were to pass before that could be achieved, and by that time the naval world was very different from that in which the 'lamps had gone out' in 1914. The specially-appointed Sub-Committee to consider the reorganisation of both the R.N.R. and the R.N.V.R. reported confidentially to the Admiralty in June 1919 but there was no implementation of its recommendations for nearly three years. This is not surprising because the Admiralty was preoccupied with more fundamental problems, and in any case large numbers of those who had fought wished to forget all about war and its horrors and to return to civilian life. It is also not surprising that few of the pre-war ratings who survived the war eventually re-engaged. This was because, firstly, the large percentage of temporary commissions (in the R.N.V.R. or in other units) which had been gained by the pre-war ratings of the Division effectively prevented further lower deck service, and, secondly, it was not to be expected that any appreciable number of those who had been interned or captured would wish to continue. So far as the pre-war officers were concerned the picture was different, and between a third and a half continued when the word was given.

If the Admiralty was for some time preoccupied it was also uncertain what to do about the Reserves. In March 1920 this led

President *during the Silver Jubilee of King George V in 1935.*

the Marquess of Graham, who had done so much to create the R.N.V.R. in 1903, to organise a gathering of R.N.V.R. officers at which the uncertainty was discussed. He described the position of the R.N.V.R. as 'unsatisfactory'. As an illustration of the 'muddle' which he claimed existed he said that one department of the Admiralty took the view that the R.N.V.R. was demobilised and discharged, while another department said that it was still serving and its liability would continue until all the Peace Treaties were ratified. He went on to say that continuance of the uncertainty would have a demoralising effect, since it made all members feel that their Service was not being taken seriously, and that if it was allowed to go on much longer many would no doubt resign. Once the R.N.V.R. ceased to exist he did not think it would be possible to resuscitate it until another war broke out, when it would be too late. Like all the other Commanding Officers of Divisions, Graham had been promoted Captain in May 1919.

To the average man in Britain in 1919 it looked as if there were no enemies capable of disturbing its future. The 'War to end wars' philosophy was taken literally, and the German Fleet was at the bottom of Scapa Flow, so it was simple for the public to demand national economy, anti-waste, disarmament and the abolition of the capital ship. All these ideas were supported in Parliament and the Press, and Naval Estimates were cut to the bone. Against this background even abolition of the R.N.V.R. could easily have been advocated as the Admiralty busied itself with the preparation of a huge Disposal List on which in May 1921 were placed 113 warships, including the *Dreadnought* herself, which were all sold for scrap. Later in the same year the remaining eight 12-inch gunned battleships, all of which were household names, suffered the same fate.

If the old Navy was largely going to the breakers some old ideas were obviously having to go as well. No sooner was Germany temporarily eliminated as a naval power than Japan arose in the struggle for sea supremacy, with the United States as a further competitor, which meant that naval rivalry shifted from the North Sea to the Pacific. Both Japan and the United States were building 16-inch battleships from 1919, and these threats to Britain's naval supremacy led her to follow suit with plans for 48,000-ton battle-cruisers with comparable armament, and even bigger battleships with 18-inch guns. The monstrous snowball had to be stopped before it had rolled too far, and finally a Conference was held at Washington in 1921–2 which agreed prescribed limits in size and numbers. It was against such a background that the Admiralty allowed the R.N.V.R. Divisions to come to life again, with effect from 1 April 1921.

Captain the Hon. Rupert Guinness, now Lord Elveden, retired as Commanding Officer of London Division on 7 June 1920 after seventeen years. His successor was Captain H. D. King, who had been promoted Captain on 1 June. He had been an officer in the Division since 1904 when he was appointed a Lieutenant, he had served in the Royal Naval Division at Gallipoli and he was Member of Parliament for Norfolk North. The appointment of Captain King clearly showed there was soon to be a new era for London Division, although it was not until May 1921 that the Admiralty formally issued new Regulations for the R.N.V.R.

A big new departure involved in these was the creation of an entirely new Special Branch, to consist of 'those officers who have no qualification for and are not required to perform executive, engineering, medical, accountant or chaplain's duties, but who are required by the Navy as specialists in certain subjects, such as civil engineers, chemists, designers, professors, artists, interpreters, etc.' As this Branch was to have green distinguishing cloth between the gold lace on the sleeve denoting rank it became known colloquially as the 'Emerald Green Navy'.[1] All this demonstrated a big step forward in naval thinking about its Reserves, at the expense of some old 'salt horse' concepts. To say that professors and chemists could be 'sailors' was a concession indeed. Another new Branch which was introduced into the R.N.V.R. was for accountant officers, who would be specially trained for coding and other similar confidential work, and who would wear the white distinguishing cloth on the sleeve which was common to all accountant officers in the Royal Navy. A number of new ranks, particularly in the Engineering Branch, were also created, adding to the general satisfaction which these innovations produced.

A provision that Warrant Rank was to be opened up to the Seaman, Signal and Telegraphist Branches of the R.N.V.R. was widely welcomed. No obligatory training was stipulated for

Warrant Officers, R.N.V.R., but candidates had to have fifteen years man's time in order to be eligible for promotion. The first such appointments of a Signal Boatswain and a Boatswain were made in August 1923 and January 1924 respectively, and both were in Sussex Division.

Another important innovation in the 1921 Regulations was the abolition of the Admiralty Volunteer Committee. In its place Admiral Commanding Reserves was put in full control of the Royal Naval Volunteer Reserve.

When it became known in July 1921 that the London County Council had granted permission for a new Admiralty training ship to be moored close to the Embankment the news soon followed that this was to be the sloop *Saxifrage* ('London Pride'), which was one of the 'Anchusa' or 'Flower' Class, collectively nicknamed by the Navy the 'herbaceous border'. She had been built by Lobnitz at Renfrew and had been commissioned in 1918 under the command of Commander E. J. Birch, R.N., as a 'Q' Ship for dealing with U boats, but she did not have time to see much real service in that capacity. It is not generally known that another sloop, the *Marjoram*, also of the 'Anchusa' Class, had originally been designated, and was about to be renamed *President*, for use as the next London R.N.V.R. drill ship, when she was wrecked on Flintstone Head while on her way to Haulbowline near Queenstown (Cork) to be fitted out. Thus only by chance did the *Saxifrage* become the London drill ship. When she was being prepared for her new role at Pembroke Dock it was reported that the existing *President* (ex-*Buzzard*) was looking strangely forlorn at her mooring, with her unstepped masts lying on her deserted deck. In September 1921 she was sold to a London firm which immediately resold her to a Dutch

President *as a 'Q' ship.*

President after her arrival in King's Reach in 1922.

firm of shipbreakers who towed her to Holland.

The new *President* (ex-*Saxifrage*) finally arrived in the Thames on 19 June 1922, and moved into the berth she has now occupied for sixty years. The large wooden deckhouse which she had been given made her resemble Noah's Ark, as the newspapers noted. Thus, with an almost brand-new *President* as the drill ship, London Division was able to resume training in earnest in September, when the summer holidays were ended.

In an endeavour to harness past glories in the solution of post-war difficulties, the King's Colour of the Drake Battalion of the Royal Naval Division was received on board the *President* at a ceremony held in that month, and laid up there.

The R.N.V.R. was unfortunate in that it had the task of rebuilding itself, on the basis of the new Regulations, at a time when the omens were unfavourable. The slaughter during the war had produced a widespread revulsion in the public mind, and this feeling lay behind the constant cry for economies in national expenditure which often had as their target the cost of the Navy. Even the pay and security of Royal Navy officers was assailed, and the 'Geddes Axe' was wielded among them with deadly effect. It is therefore not surprising that in May 1923 it was reported that in the R.N.V.R. numbers were sadly depleted and that the wastage of war had not been made good. Early in that year the strength of London Division was only 250 officers and men. Those who were responsible for the Division nevertheless set themselves the target of bringing the number up to the full permitted establishment of 1,000 by the end of 1923, by means of a vigorous recruiting campaign, but they were not

The King's Colour of the Drake Battalion being received on board President.

successful. A year later they could only claim that they had just over half the total complement.

So far as the conditions of service were concerned, a man who enrolled had to serve for four years, and make himself efficient each year. In the first year he had to attend forty drills, as of old, and in each subsequent year twenty-four drills. A great convenience to busy men was conceded in the provision that two drills could be put in on one evening. In addition to these requirements forty-two days' training in a seagoing ship had to be done during the term of service, but a longer term afloat could be served should any man desire, for while a man had to serve for a period of two weeks each year as a minimum he could remain afloat for twelve weeks if he wished and draw full pay and allowances. Yet a further attraction was that a man could go to sea at practically any time convenient for himself, which represented the great advantage he had over the Army Territorial, who had to attend camp at a specified date. At week-ends sailing parties went away in Service boats to Canvey Island and other places down the Thames.

Commodore Lord Graham presided at the R.N.V.R.'s Twenty-First Anniversary Dinner which was held in London on 15 May 1924, when Captain King of London Division proposed the toast of 'The Guests', to which Prince Arthur of Connaught, whom King George V had made an Hon. Captain in the R.N.V.R. in 1920, responded. This was the dinner at which Vice-Admiral Sir Roger Keyes said that the King of the Belgians had told him that the Royal Naval Division had 'saved the honour of the Belgian Army' by its stand at Antwerp. A month later Prince Arthur of Connaught unveiled London Division's memorial to its war dead, containing 179 names, at the entrance to the gangway to the *President* on the Victoria Embankment.

In her sixty years service as the drill ship of London Division the *President* has periodically been towed away for dry-docking and limited refitting, although the casual passer-by may never have seen her berth temporarily vacated and may presume that

RNVR
LONDON

Unveiling of the Memorial by
Captain H.R.H. Prince Arthur of Connaught,
K.G.,K.T.,P.C.,G.C.M.G.,G.C.V.O.,C.B.,R.N.V.R.
on Sunday, June 15th 1924.
at 11:30 a.m.
at H.M.S. President, Victoria Embankment, E.C.4.

Admit Bearer to Enclosure.

she is perpetually in her accustomed place. The first occasion when she was towed away was in 1926, when she was taken to Chatham Dockyard. All the bridges under which she had to pass presented difficulties, but the tightest fit of all was London Bridge, where, after the preparation of her superstructure and the lowering of her masts, there were only two feet to spare, and at the same time only two feet of water under her keel. It is only on exceptional tides that such a movement is possible for her. While the *President* is away in this fashion the opportunity is usually taken to dredge the silt from her berth. On her first absence she was away for six weeks, but as she left Chatham she was suddenly ordered to proceed to the Royal Albert Dock at East Ham to serve as a depot ship for Royal Marines during the General Strike.[2] On the way there her lower decks became flooded, but she arrived at the dock safely despite fears that she might founder. The *President* 'put to sea' again in 1929, this time for Sheerness Dockyard, and in 1933 and in 1938.

London Division suffered a tragic loss in 1930 when its Commanding Officer since 1920, now Commodore H. D. King, was drowned in a sailing accident to the yacht *Islander* off the Cornish coast. Captain N. ff. Wells was appointed to succeed him, but two years later he himself was succeeded by Cdr. J. R. Hemsted, who was called out of retirement and promoted Captain on 7 February 1933. Hemsted remained in command of the Division until after the outbreak of the Second World War, when the Division ceased to function once again, but on 13 November 1939 he was appointed to Admiralty Selection Board No. 7, serving in H.M.S. *King Alfred* at Hove, interviewing for Officer Selection. Also serving at Hove at the time in the same capacity was the former Executive Officer, Captain D. B. Nicholson. Hemsted finally retired from such duty, on age, in 1942.

The years 1930–35 included the Depression, and they proved to be a quiet time not only for the Navy generally but also for the Division. As a result, recruiting for officers and ratings was at a very low ebb. In those days executive officers were usually recruited in pairs, and only three of such joined between 1930 and 1935, when applications for entry had to be accompanied by a recommendation from some well-known person, and the first year of service was probationary. The newly-joined officers were expected to attend the *President* for two evenings each week, one of which was devoted to instruction in subjects such as gunnery and navigation and the other to duties with 'part of ship'. Seamanship came to the fore in the summer. The annual programme included a number of competitions in gun drill and sea boat drill, which were carried out on an inter-part basis. At Easter London Division usually joined Sussex Division for a weekend cruise in the gunnery cruiser attached to H.M.S. *Excellent* at Portsmouth, where 6-inch firings took place and the

Royal Naval Volunteer Reserve.
LONDON DIVISION.

Display of Drills

ON
Saturday, 6th February, 1926,
at 8.15 p.m.,
IN
H.M.S. "PRESIDENT."
Captain H. DOUGLAS KING,
C.B.E., D.S.O., V.D., R.N.V.R., M.P.,
presiding.

Admiral of the Fleet EARL JELLICOE OF SCAPA,
G.C.B., O.M., G.C.V.O., LL.D.,
has kindly consented to Distribute the Divisional Prizes during the Evening.

This Programme, Price 1/-, admits One Guest.

Seatholders are requested to be in their places 5 minutes before the Display commences, and to keep their seats while the Commanding Officer's Party is proceeding to, or leaving the Platform.

ship embarked many extra boats for pulling and sailing. Subject to weather conditions the cruise would include a visit to the Channel Islands, Torbay or Plymouth. This practical work was most valuable to both officers and ratings.

Life in the Division was not all formal work, and boxing, swimming and shooting were carried out on both an inter-part and inter-Divisional basis. Rugger, soccer and shooting at Bisley were carried out by the Division itself at its own expense. Dances were held on board the ship in order to raise money so that the Division could have its own hut at Bisley. There was also sailing on the river at weekends at Long Reach, when whalers would get a tow down by lightermen on a Friday evening and a return tow on the Sunday. There was also a flourishing amateur theatrical party which performed for a week in the spring, and in the early Thirties there was a series of musical productions, written by Lieut. Dundas Grant, which were set in Ruritania in the style of the *Merry Widow*.

The Thirties saw London Division on show to the public more than in any previous era, through a series of ambitious Navy Weeks. The second of these, in 1932, involved a week-long programme of nightly Displays on board the *President*, in which, at nine o'clock each evening, an 'attack' was made on the ship by 'hostile' forces represented by her own motor boats, while her crew went to night defence stations, dowsing all lights and manning the guns and search-lights. Suspicious craft moving up and down the river were challenged by lamp with the Morse signal for the night, and 'fire' was opened as necessary, with reduced charges. The bridge of the ship was floodlit so that all

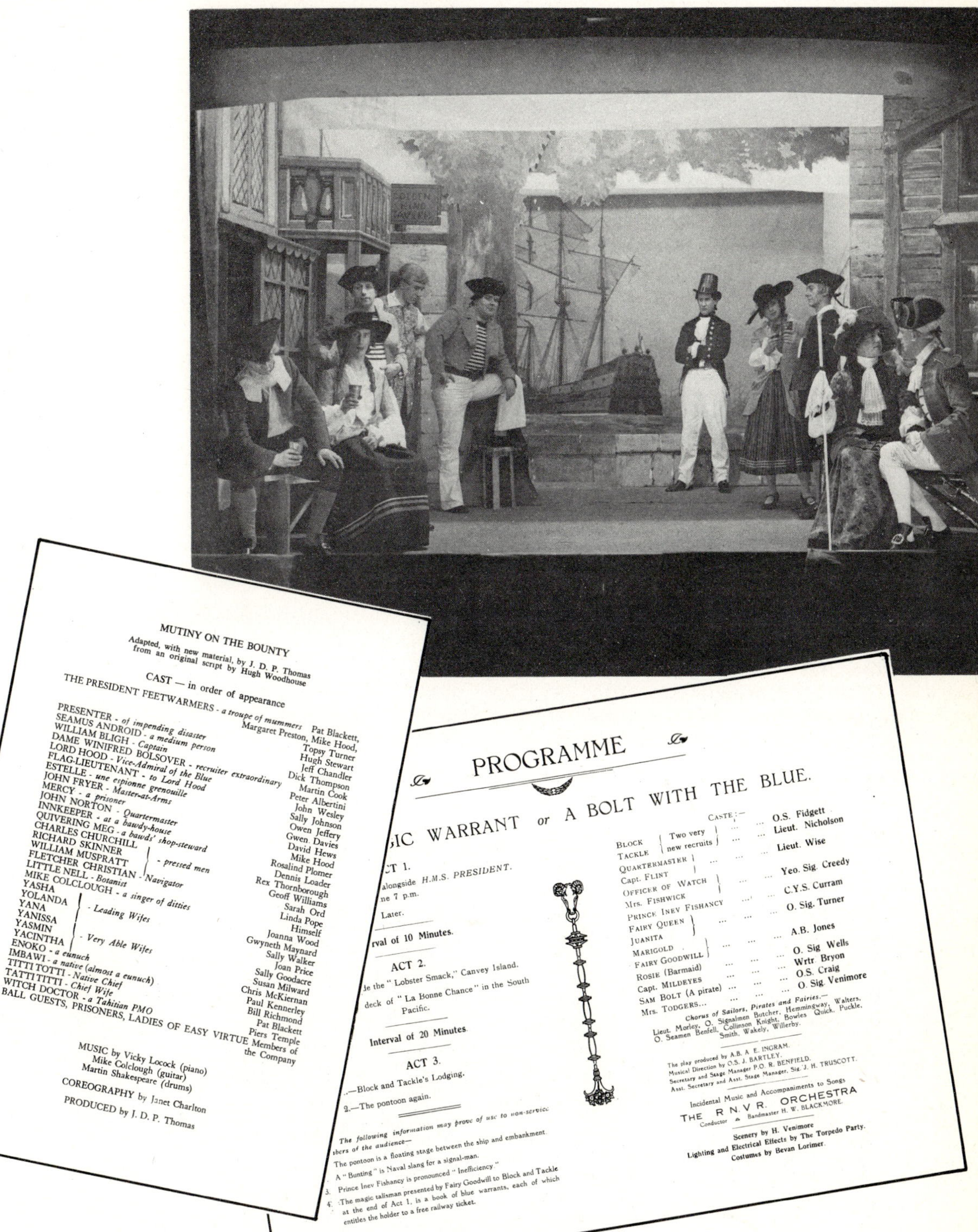

MUTINY ON THE BOUNTY

Adapted, with new material, by J. D. P. Thomas
from an original script by Hugh Woodhouse

CAST — in order of appearance

THE PRESIDENT FEETWARMERS - *a troupe of mummers*
PRESENTER - *of impending disaster*
SEAMUS ANDROID - *a medium person*
WILLIAM BLIGH - *Captain*
DAME WINIFRED BOLSOVER - *recruiter extraordinary*
LORD HOOD - *Vice-Admiral of the Blue*
FLAG-LIEUTENANT - *to Lord Hood*
ESTELLE - *une espionne grenouille*
JOHN FRYER - *Master-at-Arms*
MERCY - *a prisoner*
JOHN NORTON - *Quartermaster*
INNKEEPER - *at a bawdy-house*
QUIVERING MEG - *a bawds' shop-steward*
CHARLES CHURCHILL / RICHARD SKINNER / WILLIAM MUSPRATT - *pressed men*
FLETCHER CHRISTIAN - *Navigator*
LITTLE NELL - *Botanist*
MIKE COLCLOUGH - *a singer of ditties*
YASHA
YOLANDA / YANA / YANISSA - *Leading Wifes*
YASMIN / YACINTHA - *Very Able Wifes*
ENOKO - *a eunuch*
IMBAWI - *a native (almost a eunuch)*
TITTI TOTTI - *Native Chief*
TATTI TITTI - *Chief Wife*
WITCH DOCTOR - *a Tahitian PMO*
BALL GUESTS, PRISONERS, LADIES OF EASY VIRTUE

Pat Blackett, Margaret Preston, Mike Hood,
Topsy Turner
Hugh Stewart
Jeff Chandler
Dick Thompson
Martin Cook
Peter Albertini
John Wesley
Sally Johnson
Owen Jeffery
Gwen. Davies
David Hews
Mike Hood
Rosalind Plomer
Dennis Loader
Rex Thornborough
Geoff Williams
Sarah Ord
Linda Pope
Himself
Joanna Wood
Gwyneth Maynard
Sally Walker
Joan Price
Sally Goodacre
Susan Milward
Chris McKiernan
Paul Kennerley
Bill Richmond
Pat Blackett
Piers Temple
Members of the Company

MUSIC by Vicky Locock (piano)
Mike Colclough (guitar)
Martin Shakespeare (drums)

COREOGRAPHY by Janet Charlton

PRODUCED by J. D. P. Thomas

PROGRAMME

…GIC WARRANT *or* A BOLT WITH THE BLUE.

…CT 1.

…alongside *H.M.S. PRESIDENT.*

…ne 7 p.m.

… Later.

…rval of 10 Minutes.

ACT 2.

…de the " Lobster Smack," Canvey Island.

… deck of " La Bonne Chance " in the South Pacific.

Interval of 20 Minutes.

ACT 3.

…—Block and Tackle's Lodging.

2.—The pontoon again.

The following information may prove of use to non-servic… …bers of the audience—

… The pontoon is a floating stage between the ship and embankment.
… A " Bunting " is Naval slang for a signal-man.
3. Prince Inev Fishancy is pronounced " Inefficiency."
4. The magic talisman presented by Fairy Goodwill to Block and Tackle at the end of Act 1, is a book of blue warrants, each of which entitles the holder to a free railway ticket.

CASTE:—

BLOCK / TACKLE { Two very new recruits } … … O.S. Fidgett / Lieut. Nicholson
QUARTERMASTER / Capt. FLINT … … … Lieut. Wise
OFFICER OF WATCH / Mrs. FISHWICK … … Yeo. Sig. Creedy
PRINCE INEV FISHANCY … … C.Y.S. Curram
FAIRY QUEEN / JUANITA … … … O. Sig. Turner
MARIGOLD / FAIRY GOODWILL … … … A.B. Jones
ROSIE (Barmaid) … … … O. Sig Wells
Capt. MILDEYES … … … Wrtr. Bryon
SAM BOLT (A pirate) … … … O.S. Craig
Mrs. TODGERS… … … … O. Sig. Venimore

Chorus of Sailors, Pirates and Fairies.—
Lieut. Morley, O. Signalmen Butcher, Hemmingway, Walters, O. Seamen Benfell, Collinson Knight, Bowles Quick, Puckle, Smith, Wakely, Willerby.

The play produced by A.B. A. E. INGRAM.
Musical Direction by O.S. J. BARTLEY.
Secretary and Stage Manager P.O. R. BENFIELD.
Asst. Secretary and Asst. Stage Manager, Sig. J. H. TRUSCOTT.

Incidental Music and Accompaniments to Songs
THE R.N.V.R. ORCHESTRA
Conductor … Bandmaster H. W. BLACKMORE.

Scenery by H. Venimore
Lighting and Electrical Effects by The Torpedo Party.
Costumes by Bevan Lorimer.

Concert Party Programmes.

movements of the officers controlling the 'action', the searchlight crews and the upper deck guns' crews were in plain view of the people on the Embankment and on Blackfriars Bridge.

In these Navy Weeks, which were a regular annual feature from 1931 to 1938, the *President* was open for inspection by the public, each year in late August or early September. They were organised entirely by the officers and men of the Division, and all profits were devoted to R.N.V.R. charities. Londoners took a great interest in the events, and in 1937 over 20,000 people visited the ship, where the Seamen's Band from the Royal Naval Barracks at Chatham played on board daily, and the ceremony of 'Sunset' was carried out at eight o'clock each evening. A principal feature that year was a panorama of the Coronation Review with over 150 ship models on a scale of 150 feet to the inch. Even greater numbers attended for the Navy Week held in 1938, when the officers, petty officers and men took care to explain in the programme that the reason why they were unable personally to welcome their guests and show them round the ship during the afternoons was because they had to follow their work or business during the day, when the responsibility fell on the commissioning party and the permanent staff. A charge of sixpence was made for admission on board. In reporting the Navy Week of 1934 the *Daily Express* said: 'They have gladioli on the mess deck this week. Yes, gladioli. And maiden-hair fern, and drawing room pastries, and tea at 1½d. a cup.'

The only serious dissenter on these busy occasions seems to have been Peter, the *President's* overweight tabby cat, only one of whose ears was able to stand upright and who was officially borne on the ship's books for a one shilling and sixpence a week victualling allowance for his milk and cat's meat. In Navy

Cleaning guns on board President *1936 in best rig for Navy Week, 1936.*

The Lord Mayor, Sir Stephen Killick, inspecting the Guard of Honour before opening London's Navy Week. The officer of the guard is Lieut. H. Sketch.

Weeks he betook himself off below and hid away from visitors. Peter became ill in 1937 and though Cdr. Withrington took him to an animals' nursing home it was not possible to save him.

According to custom a prize-giving was held in the *President* each year before Christmas, and many well-known people were entertained on board beforehand, including the then Prince of Wales, the Duke of Kent and the Duke of Gloucester, who on one occasion ordered the Gunnery Shield to be filled with beer and insisted on each officer trying to drain it. He then ensured nobody succeeded by 'accidentally' jerking the Shield at the critical moment.

In May 1935 King George V and Queen Mary celebrated their Silver Jubilee, and London Division lined the Embankment abreast the ship on a lovely summer's day. When the *President* fired a Twenty-one Gun Royal Salute as the King and Queen drove along the Embankment after attending the Thanksgiving Service at St. Paul's Cathedral, it was the first time the Navy had saluted the King within the sound of Bow Bells.[3] Later, at the Silver Jubilee Review at Spithead on 16th July 1935, a large contingent from London Division was among the 3,000 Reservists who helped to man the ships of the Reserve Fleet. When the King died in the following January, London Division lined Whitehall opposite the Horse Guards as his funeral procession went by.

At the end of 1936 King Edward VIII abdicated the throne. On the evening of the abdication, Commander Withrington and

Royal

18·xii·34

Edward P

The Prince of Wales

The Duke of York inspecting the guard provided by London Division at the R.N.V.R. Association Parade in May 1936.

The Duke of Kent at the annual prizegiving only six days after the Abdication.

George.

17th December. 1936.

Brothers

December 14th 1937.

Henry.

The Duke of Gloucester.

Members of London Division lining the processional route in Whitehall for the Coronation of King George VI in May 1937. (Note the double lining.)

the wardroom officers, after the loyal toast to the new King, in a spontaneous gesture drank the health of their former Sovereign in the Mess. Six months later, when King George VI was crowned in Westminster Abbey, London Division once again lined Whitehall, this time on both sides near the Cenotaph. The Home Fleet, which had come up the river and was anchored off Southend, detached the destroyer *Kempenfelt* which came up to London and anchored in the Pool. When Captain Hemsted was rowed down to her in a whaler to call on her Captain the tide was ebbing strongly, and on his return to the *President* it took him twenty minutes to get under London Bridge.

The Admiralty announced in August 1937 that the *Chrysanthemum*, a sister ship of the *President*, was to be converted to serve as an overflow ship and signal school moored upstream of the *President*. She had been built in 1917 and had just ended service in the Mediterranean as a fleet target and photographic ship. It was first expected that she would arrive within a few weeks, but it was not until 12 May 1939 that she finally came alongside the Embankment after fitting-out at Portsmouth. In her more active days when she was towing targets for the big ships' practices one of the battleships put the deflection on the

Characters of the 1930's.

wrong side and sent a 15-inch shell, luckily a dummy, into her. Her captain soon afterwards had to submit the usual quarterly return, and he answered the question as to the general behaviour of the ship's company with: 'My ship's company behaved with great coolness while under fire from H.M.S. *Barham*'. As the *Barham* was Acting Flagship at the time, and so received the report, the Staff were not pleased.[4]

During 1937 and 1938 recruiting took place on board the *President* for the newly-established Royal Naval Volunteer Supplementary Reserve, which in its first four months of operation reached a strength of over 1,000. All the new volunteers had some experience of seafaring, many of them owning yachts and other small craft, and in emergency they were pledged to report for duty on being called upon after mobilisation. Originally it

144 THE SPHERE APRIL 25, 1936

CITIZEN SAILORS TAKE TO THE SEA

Men of the Royal Naval Volunteer Reserve Spend Their Easter on a Warship

RUM ROW: Serving out the daily ration of grog at 1 bell (lunch hour) aboard H.M.S. *Curacoa*. In accordance with tradition and with Admiralty Regulations a sergeant of Marines superintends

LADS OF THE "WAVY NAVY": Men of the Royal Naval Volunteer Reserve cleaning a gun on the 4,000-ton cruiser *Curacoa* aboard which they spent the four days of the Easter holiday, gaining invaluable practical experience to fit them for service in the event of a national emergency. As a rule they have scant opportunities for going to sea, their civil occupations and responsibilities holding them securely to the land

SEMAPHORE CLASS ON THE FORECASTLE: The R.N.V.R. cadets practising signalling while the *Curacoa* lay at anchor off the coast. Not a moment of the few short days afloat was wasted and (*right*) the cadets are seen undergoing sailing instruction. The *Curacoa* belongs to the *Ceres* class and was completed during the war. Her armament includes five 6 in. guns

NOT all of us would give up our Easter holidays in order to make ourselves better fitted to serve the Empire in time of need. Yet this is what sixty-eight men of London and forty-one men of Sussex did this last Easter.

They were officers and men of the "Wavy Navy," as the Royal Naval Volunteer Reserve is affectionately called—because the rank stripes of its officers are wavy. They left their varied civil tasks at the usual time on Thursday evening. At eleven o'clock that night a tug put them on board the cruiser *Curacoa* at Spithead. A few minutes later they were in a warship under way, and hard at work.

Officers and men of the Royal Naval Volunteer Reserve have scant opportunities of going to sea. Their civil occupations and responsibilities hold them to the land week after week and month after month. Thus, when an opportunity occurs there is no time to be lost. H.M.S. *Curacoa* had not rounded the Nab before some of these citizen sailors were on watch, either as look-outs, assistant quartermasters, or assistants to the officer of the watch.

And so it was from Thursday night until the evening of Easter Monday, when the seventeen officers and ninety-two ratings of the R.N.V.R. departed to pick up the threads of their lives "on the beach."

Throughout the "holiday cruise" there was not a moment wasted. With the ship at sea all day the reservists were given practice in the handling of the ship, at picking up buoys dropped overboard, and at firing the 6-in. guns. In harbour there was not a moment of daylight in which boats were not away either under oars or under sail.

And because the R.N.V.R. realises that full training value can best be achieved when there is an element of competition, there were boat-races whenever the ship was in harbour, and even when the ship was at sea. The latter took the form of a "man-overboard-race." Both life-buoys were dropped, and both sea-boats lowered. The men of London manned one of the sea-boats, the men of Sussex the other, and they raced to pick up the buoys and then get back to the ship, hooked on and hoisted. London won, but Sussex had its revenge with the whaler-race held in Plymouth Sound. This race is an annual event for a trophy consisting of a gilded wooden Easter egg mounted in a wooden egg-cup. It is always raced for on Easter Day, and never was trophy of gold or silver more keenly contested.

Signalling was not forgotten. Wireless experts carried out exercises with H.M.S. *President*, the R.N.V.R. drill ship moored in the Thames at Blackfriars, while signalmen were exercised with flashing lamps and in "flag-wagging."

One of these signalmen of the R.N.V.R. was a London medical student. This was only one of many walks of life from which these citizen sailors came. There were omnibus conductors, tram conductors, ice-cream vendors, shop assistants, employees of gas works and factories, and fishermen. There was even a solicitor who, when not wearing the uniform of an ordinary seaman, conducts his own practice.

ALL these men, called to the sea by instinct and the desire to fit themselves for the service of the country in emergency not only give up their Easter holiday to this training. They give up a certain number of evenings to drill after the normal work of the day is over. There is no question of monetary gain commensurate with the time spent upon training. Naturally, when at sea they are paid just as active service ratings are paid.

For this particular cruise, which is an annual event but which this year has given training to 25 per cent more R.N.V.R. ratings than previous cruises, ratings are selected from those who, although enrolled in the Royal Naval Volunteer Reserve, have not yet had the opportunity of serving at sea in a warship. That there is great competition for these sea-training cruises is shown by the fact that, for the cruise which has just ended, fifty ratings had to be selected from the London Division of the R.N.V.R. out of 130 who volunteered.

The pity of it is that shortage of ships and the necessity for economy have made it impossible for similar cruises to be carried out for the benefit of the R.N.V.R. divisions stationed in the West and North of England and in Northern Ireland.

Kenneth Edwards

was not intended to provide for them any official training facilities, but the immense enthusiasm displayed by these volunteers impressed the Admiralty, which finally arranged for them to make short training cruises in units of the Royal Navy. The Royal Naval Volunteer Supplementary Reserve was different from the permanent R.N.V.R. in that in time of peace its members held no rank, wore no uniform and drew no emoluments, and in time of war would be given only temporary commissions. Certain London members of the R.N.V.S.R. formed themselves into the 'London Flotilla', in order to provide training at their own expense, purchasing out of their own resources two 56-ft. ex-Naval steam picket-boats and maintaining them within reach of London for the use of their fellow R.N.V.S.R.'s.

The Munich Crisis at the end of September 1938 led to a great increase in volunteers for the permanent R.N.V.R., but the *Chrysanthemum* could not be there in time to assist in dealing with them. The Crisis made many Londoners realise for the first time that they should belong to some auxiliary unit. Following the call-up of Naval Reserves the pressure on the *President* became so great that on 28 September a notice chalked on a blackboard at the entrance to her gangway read: 'R.N.V.R. IT IS REGRETTED THAT NO MORE RECRUITS CAN BE ACCEPTED FOR THE ABOVE. IN THE EVENT OF HOSTILITIES APPLICATION TO JOIN ANY OF THE NAVAL SERVICES SHOULD BE MADE TO 85 WHITEHALL.' The *President* was sent away to Sheerness for dry-docking and some refitting on 14 November, and she did not return to her mooring until 18 January 1939, having been fitted with two new 4-inch Anti-Aircraft guns on the upper deck and having briefly stuck under Blackfriars Bridge. The presence of these new guns did much to allay the widespread fear that members of the Division would yet again have to fight on land. This fear was not wholly unfounded, as later research has revealed that active consideration was being given at that time to the formation of yet another Royal Naval Division.

The Naval Estimates of March 1939 revealed a large increase in the establishment of the R.N.V.R. for the introduction of Anti-Aircraft training, so as to provide guns' crews for ships in Reserve. To enable this, some new Divisions were created, including the Humber Division at Hull, where the Reserve Fleet cruiser *Calcutta* became the drill ship after fitting-out for her new duty. This training was under way by June, when it was announced that certain old 'V' and 'W' Class destroyers were allocated to R.N.V.R. divisions, but in London the *President* herself, with the new A.A. guns and an H.A./L.A. Director in the bridge structure, had been suitably equipped for such purposes. The training started which would enable London Division to man the armament of four 'C' Class cruisers which had been converted for Anti-Aircraft duties with the Fleet. Four A.A. units were trained by August, each unit comprising three

Gunnery practice on board the President.

officers and fifty ratings. Lieut. Whittow's unit went to the *Coventry*, Lieut. Burns' to the *Curlew*, Lieut. MacGregor's to the *Cairo* and Lieut. Noble's to the *Calcutta*. All these units were at sea by the end of the month, by which time, once again, the Fleet had been mobilised. There were now so many signal ratings in the division that the Council of the Stock Exchange kindly allowed the House to be used as a signal school in the evenings. The basement of Temple Chambers was taken over as classrooms.

Another important branch of naval signalling was Wireless Telegraphy, and in that also the R.N.V.R. was playing its part. In the days just before the Second World War W.T. exercises were regularly carried out between the *President* and the other

divisions, and from time to time unaccountable 'black outs' occurred, to the mystification of all the experts, who considered they had taken into consideration all the possibilities. However, the one thing they had overlooked was the existence of the double tramway tracks which up to the early Fifties came on to the Embankment from Blackfriars Bridge and left it over Westminster Bridge. The situation is difficult to appreciate for someone who only knows the Embankment in its modern form, but in the days before, during and just after the Second World War these tracks ran very close to the pavement on the riverside, and as there were no overhead trolley-wires the current was picked up from a central conduit or slot along which each tram drew a heavy 'shoe'. Eventually the simple truth dawned – as the tide in the river fell, the aerials on the ships were level with or below these central conduits in the roadway between the tracks and were effectively screened by them, thereby preventing W.T. communication.

In the rearmament rush during 1938/9 the Services put in special telephone and telex lines to link up all Service establishments; it was called the Defence Telecommunications Network (D.T.N.). The Admiralty turned to London Division for help. Commissioned Signal Bosun Peter Curram (the only C.S.B., R.N.V.R.,), whose shore-side job was Postmaster at the House of Commons, was appointed by Admiral Commanding Reserves to recruit and organise Reservists to run the Navy's D.T.N. Switchboard.

On 27 August 1939 the Royal Proclamation was issued which mobilised the Royal Naval Reserves, including the R.N.V.R., for service with the Fleet. By now, many members of the Division were already afloat, so when on 3 September the signal 'Winston is back' was flashed to the Fleet, the R.N.V.R. was able to rejoice with everyone else, happy in the knowledge that he would not again use them as soldiers.

CHAPTER FIVE

'Down to the Sea in Ships'

With the outbreak of the Second World War London Division once again ceased to have a corporate Thames-side existence, although the Commanding Officer, Captain J. R. Hemsted, remained in command in the *President* for about two months. The history of the Division during the period of hostilities may only be traced therefore in the individual experiences of its mobilised members, wherever they happened to serve. It was a different story from that of 1914–18, for, whereas in 1914 about 75% of the officers and men of the Division had gone into the Royal Naval Division to fight on land, in 1939 the Admiralty, which had learnt some salutary lessons concerning its Reserves, showed better husbandry in the use of the R.N.V.R. The Division's mobilisation records show that close on a thousand men had been drafted away, chiefly to the three Royal Naval Barracks, for further disposal to ships and establishments, even before war was declared on 3 September 1939. Excluding those declared medically unfit, approximately 1600 men from the Division were eventually sent for service with the Royal Navy, to make, with the 300 officers, approximately 1900 in all.

Over 200 of these London men went to war by manning the four anti-aircraft cruisers to which reference has already been made. To some extent, therefore, the corporate nature of London Division continued even after the declaration of war, and thus its history may be specially traced in the fortunes of the four ships concerned – the 'Ceres' class *Coventry* and *Curlew*, and the 'Capetown' class *Cairo* and *Calcutta*, each of which had been laid down during the First World War and converted to its new A.A. role in the years immediately preceding the Second.

In the *President* during the spring of 1939 the Fo'c'sle (FX), Foretop (FT), Maintop (MT) and Quarter Deck (QD) subdivisions of her ship's company came to be seen as the respective units which would be sent to these four cruisers on or before mobilisation, when the time came. They trained hard, sometimes with the assistance of R.A.F. aircraft flying high overhead, and the arrival of the additional drill ship *Chrysanthemum* on the

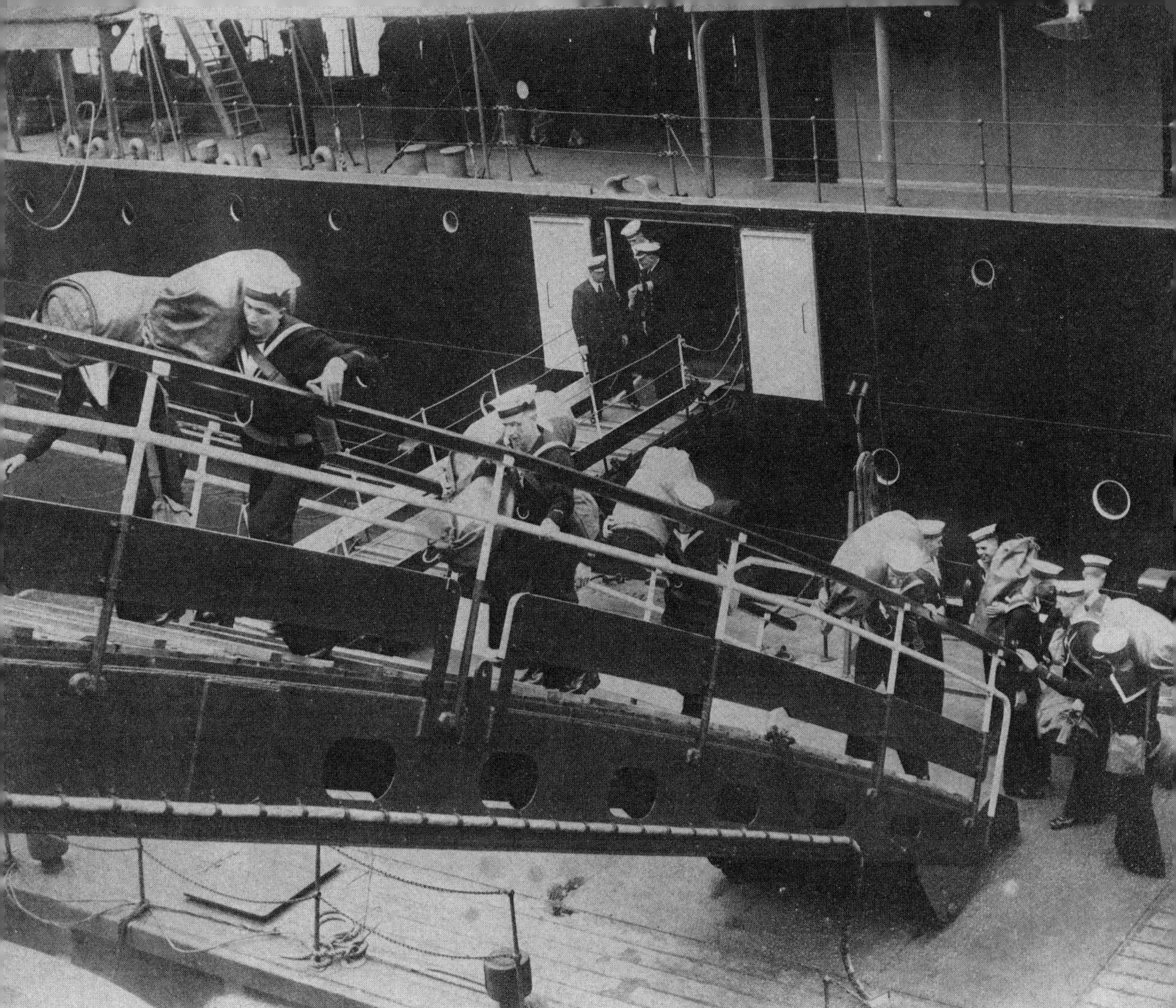

Members of the Division mobilising for duty with the Reserve Fleet 31 July 1939.

12 May 1939 enabled the *President* to be devoted almost exclusively to the work of gunnery training.

The FX unit (54 men) joined the *Coventry* at Portsmouth on 20 August, the FT unit (57 men) and the QD unit (52 men) joined the *Cairo* at Portsmouth and the *Calcutta* at Hull respectively on 25 August, while the MT unit (54 men) joined the *Curlew* at Chatham on 31 August. The officers who accompanied each unit to sea were, in the *Coventry*, Lieut. H. S. Whittow and Lieut. J. S. Elliott; in the *Cairo* Lieut.-Cdr. A. R. McDougall, Lieut. I. L. MacGregor, Sub-Lieut. N. N. J. Bodger and Sub-Lieut. B. H. Moss; in the *Calcutta* Lieut. C. P. C. Noble, Sub-Lieut. L. G. Hinge and Mid. A. F. Hewlett; and in the *Curlew* Lieut. N. McG. Burns, Sub-Lieut. H. N. Hinge and Mid. A. M. L. Merry[1]. Thus 226 officers and men of London Division, all of whom in view of their training saw themselves as gunnery specialists, were afloat in H.M. Ships some days before the declaration of war.

On 21 October 1939 the *Calcutta* was attacked in the North Sea by waves of Heinkels, which were driven off with casualties

Off to war.

to the raiders and little harm to the convoy she was escorting. That night the Commanding Officer, Captain H. A. Packer, R.N., wrote to his wife Joy (as she told later in a book): 'You'll have heard of our adventures over the wireless. It's been a wonderful day, really most exciting, with things happening all the time. My inexperienced crew were magnificent, jumping about like cats to man the guns. It was a grand way of spending Trafalgar Day.'

All the four cruisers were at sea continuously throughout the bitter winter of 1939–40 and participated in the Norwegian Campaign. While approaching the Lofoten Islands on 7 May 1940 in company with the aircraft-carriers *Ark Royal* and *Glorious*, the *Curlew* suffered a serious accident through a defective firing circuit in her pom-pom director, and this caused the deaths of five men while gun-barrels were being cleaned. Two of the dead, Able Seaman P. Lamsdale and Able Seaman J. S. Jenkins, belonged to London Division. Three weeks later she was hit by German bombs while endeavouring to protect troops who were labouring to build an airstrip at Skaanland in Lavangs Fjord close to Ofotfjord and Narvik. She ran on to a reef while out of control and sank within two hours, and four officers, including Sub-Lieut. H. N. Hinge and Sub-Lieut. A. M. L. Merry of London Division, and five ratings were lost.

A large number of the *Curlew's* survivors were drafted to the new Cruiser *Naiad*, in which, after qualifying at Whale Island, Lieut. Burns, of London Division, became Gunnery Officer. The

Naiad was sunk on 10 March 1942, so former *Curlew* survivors became survivors again, but with only one London Division casualty.

In the assault on Narvik on 28 May 1940 the *Cairo*, in which Admiral of the Fleet the Earl of Cork and Orrery flew his flag, was carrying the naval and military commanders and their staffs, and wearing battle ensigns, the Union Flag, the French Tricolour and the Polish and Norwegian national flags. Thus arrayed, she was easily identified as a prime target in Ofotfjord for the German bombers. She was hit by two bombs and suffered nine killed and twenty-one wounded, including Sub-Lieut. Moss, who had a foot almost severed, and Able Seaman E. G. Marsh, who died of his wounds[2]. The *Coventry*, which also was in Ofotfjord, received her share of bombs, a splinter from one of which killed Able Seaman A. Ballard of London Division. The *Calcutta*, which had been involved in the operations at Andalsnes, Namsos, Mo and Skjel Fjord, and which had suffered many hours of air attacks, had been ordered back to Rosyth in mid-May and was then sent to the Kent coast to stand by for use during the emergency facing the British Expeditionary Force in France.

Operation 'Dynamo', the name given to what was to become known as the Dunkirk Evacuation, began on 26 May, and shortly before midnight on the 27th the *Calcutta* sailed with six destroyers to embark all possible British troops from the beach at La Panne, three miles east of Dunkirk. Her own boats and other craft made repeated trips from the beach to the ship after she anchored at 0250. The work continued as darkness gave way to dawn and the German bombers arrived to bomb and strafe the beach and the ships. The *Calcutta*'s guns offered a fierce barrage as the weary soldiers were brought alongside. She left for the dash home to the Thames Estuary at 0712 carrying 607 soldiers. On the following evening she went across to La Panne again and returned this time with 1,000, having remained till early afternoon. On subsequent days she was used for purposes of giving A.A. cover in mid-Channel to the hard-pressed and overloaded ships as they returned from the beaches. The *Calcutta* came through the operation unscathed. As late as 24 June she left the roadstead at St. Jean-de-Luz with destroyers escorting

London's Navy at Waterloo Station 24 August 1939, on its way to war.

troopships carrying a Polish battalion and some French and British elements to Britain. That night, while manoeuvring, the *Calcutta* collided with the Canadian destroyer *Fraser* at twenty-five knots and cut her in half with heavy loss of life, but all the other troopships and warships arrived home safely.

While the Dunkirk drama was unfolding, the faithful *Coventry* was still operating in the Norwegian fjords escorting returning troopships, and she narrowly missed the German heavy units which sank the aircraft-carrier *Glorious* not many miles away from her.

By the end of June 1940 the officers and men of London Division who had been manning these 'C' class cruisers could fairly claim that they had been 'blooded' in the art of naval warfare and had experienced no small baptism of fire from the enemy. Alas, the *Curlew* was no more, but the other three were yet to gain twelve more battle honours between them, virtually all in the Mediterranean.

The *Calcutta* and the *Coventry* went out there in August 1940 in a powerful convoy destined for the reinforcement of Malta, now beleaguered by the collapse of France and the entry of Italy into the war. Throughout a period of almost a year these ships were frequently in company, providing A.A. defence for Malta convoys, for troopships to Greece and from Greece to Crete, or in support of the Desert Army, and they were often under air attack.

On 13 December the *Coventry* was torpedoed off Sidi Barrani by an Italian submarine, but got safely back to Alexandria where her ship's company were able to enjoy Christmas leave. During the defence of Crete in May 1941 the *Coventry* offered a most spirited and successful defence of the crowded hospital ship *Aba* which was being attacked by Stuka dive-bombers, in an action in which the *Coventry's* director-layer won the Victoria Cross and Able Seaman S. Fisher, of London Division, who was in the director with him, was wounded by the concentrated machine-gun fire it received.

On the morning of 1 June 1941, as the *Calcutta* and the *Coventry* advanced to meet the cruisers returning with the last soldiers to be evacuated from Sphakia in Crete, the *Calcutta* was attacked and sunk 100 miles north of Alexandria by bombs from a Junkers 88, going down within five minutes. Three officers and 116 ratings were lost, and these included four men from the Division, Able Seaman K. Anderson, Able Seaman H. Gardner, Able Seaman J. Stanley and Able Seaman L. Williams. Leading Seaman J. Mercer of the Division was wounded. The *Coventry* rescued 256 survivors and took them to Alexandria at twenty-nine knots. Later, the C-in-C. stated that the *Calcutta* 'had a first-rate ship's company'.

When the *Cairo's* bomb damage received off Norway had been repaired she was bedevilled by engine-room and other breakdowns which dulled much of the lustre she had begun to show in Ofotfjord. Although she was escorting convoy WS 8B in the Atlantic when the *Bismarck* was found to be at large, for about eighteen months her story was generally one of routine Atlantic convoy arrival and departure escort duties while working from the Clyde. Early in 1942 she went with the cruiser *Nigeria* to Murmansk in North Russia, carrying members of the Supreme Soviet Command who had been visiting Britain. Two quick visits to the Mediterranean followed, protecting the U.S.S. *Wasp*. To the *Cairo*'s ship's company the sunshine was delectable, but when she was given a third opportunity to experience it the purpose was to take part in Operation 'Harpoon', in which her task was to cover a convoy which would supply Malta from the west, in concert with Operation 'Vigorous' which would try to do the same from the east with the A.A. protection of the *Coventry* from Alexandria. In a nice poetic way, old but long-separated London Division shipmates in the two cruisers would share in a joint enterprise. Above all, the *Cairo* was to be back in the A.A. defence business at its grimmest.

In Operation 'Harpoon' it proved to be warm work. The *Cairo* led the close escort and went right through to Malta with destroyers. Her convoy from the west, which was supported by bigger ships, suffered seven air attacks, and when near Pantellaria she and the destroyers fought against a squadron of Italian surface ships consisting of two cruisers and five destroyers. She covered her convoy with smoke and then joined in the surface action. She was hit by two shells. The convoy from the west lost four ships, but two got safely through to Malta. Two destroyers of the escort were sunk.

The convoy from the east, Operation 'Vigorous', containing the *Coventry*, was even more powerfully supported by bigger ships, and was commanded by Rear-Admiral Sir Philip Vian. In his words it experienced 'all known forms of attack', including the Italian battleships *Vittorio Veneto* and *Littorio* and E-boats. The air attacks became almost continuous, and when only one-third of the ships' ammunition remained the C-in-C. recalled it to Alexandria. Operation 'Vigorous' had failed, and it delivered no merchant ships to Malta.

Though Britain's naval and military fortunes in the Mediterranean theatre were in the balance at that moment, as was the fate of Malta, plans were made for another Malta convoy in August in which three operational carriers, two battleships, six cruisers, the *Cairo* and twenty-four destroyers were to take part. The *Furious* was also to go (to fly-off thirty-eight fighters to Malta) and fourteen fast merchantmen. It was to be a big movement, and it received the name Operation 'Pedestal'. With the Arctic convoys suspended after PQ 17 in July, 'Pedestal', for which Home Fleet ships could now be spared, was in effect a naval demonstration after the Arctic

Coventry, *her AA guns manned by London Division, pictured in home waters 1940.*

disaster, and members of London Division were there in the *Cairo*.

Three U-boat, two E-boat and seven air attacks by no less than 263 planes were hurled against the 'Pedestal' ships, which passed Gibraltar on 10 August, and they caused serious losses, including the *Cairo*. She was hit at 2000 on 12 August by a torpedo in a salvo of four fired from the Italian submarine *Axum* near the Skerki Bank. The other three also found targets – the cruiser *Nigeria* and the tanker *Ohio* (two). The *Cairo* was at the vortex of all this, and at the same time waves of torpedoes were coming from German and Italian planes. The sea and sky were red with flame, yet the *Cairo*'s guns still fired even after she had stopped. When 'Abandon ship' was ordered in the *Cairo* the men clambered over the guardrails on to the destroyer *Wilton* which had come alongside the starboard waist. Then the *Pathfinder* gave the *Cairo* the *coup de grâce* with torpedoes, aided by the guns of the *Derwent*. Twenty-three men in the *Cairo* were killed by the Italian submarine's torpedo, but all her London Division men were among the 377 who survived and returned to Gibraltar in the *Wilton*, to be brought home from there by the *Rodney*. Five of the merchant ships reached Malta, and these included the *Ohio*, whose master was awarded the George Cross.

Cairo at sea, August 1940.

With the *Cairo* gone, only the *Coventry* remained of the four cruisers which London Division had manned. On 25 August her R.N.V.Rs. celebrated the completion of three years since they joined her in 1939, during which time less than ten of the original draft had left the ship. The desperate Mediterranean situation in the late summer of 1942, however, made many of them wonder how much longer the *Coventry* could survive. G. G. Connell, who served in her and who wrote the story of the four cruisers in his *Valiant Quartet*, has told with much perception of the feelings of his shipmates at that time, and of how at Port Said on 12 September their suspicions that, ashore, security had been breached were confirmed when *felucca* men who came alongside with their laundry informed them that officers' luggage was to go to the Fleet Club and the ship was going to Tobruk. Later that day the *Coventry* sailed with four 'Hunt' class destroyers, to rendezvous with the 'Tribal' class *Sikh* and *Zulu* as well as four more 'Hunts' carrying 350 Royal Marines.

The *felucca* men were right about Tobruk. That night, a landing was attempted, but things went badly wrong. The *Coventry* and her destroyers had turned back when the two 'Tribals' proceeded inshore off Tobruk to support the landing, in order that she could provide A.A. cover as they returned to Alexandria in daylight. However, the *Sikh*, revealed by searchlights, was attacked by shore guns, hit and sunk. The *Zulu* tried to assist but was ordered away, and the *Coventry* went to meet her to give her cover. Before she could do this, the *Coventry* was attacked by fifteen Stukas and was badly hit by bombs. Her bow was blown off back to 'A' gun and a crater was blasted through her. When 'Abandon ship' was ordered, the *Beaufort* came alongside and took off some survivors, while the *Dulverton* rescued others who were in the water. When only the dead remained on board the *Coventry* the Hunts tried to sink her by gunfire, for the *Zulu*, still to the westward, had to be assisted. As they were firing, the *Zulu*, under heavy attack from the air, came over the horizon, and the Hunts turned towards her, to aid her defence. Despite her own preoccupations, the *Zulu* approached the *Coventry* and put torpedoes into her, sending her down with her dead. The *Zulu*, alas, did not escape a further Stuka attack and was sunk.

The whole operation against Tobruk had been a disaster. Three officers and sixty-one men of the *Coventry* had lost their lives, including two men from London Division, Able Seaman G. C. Ball and Able Seaman G. Sullivan, known as 'Spike', who before the war had been a chef at Stone's Chop House in London and had represented the Division at boxing and rowing. In the *Coventry* he had acted as Officers' Steward, and he became an Oerlikon gunner. He was firing as the *Coventry* came under attack, and he was killed at his gun. Some of the survivors were drafted to other ships of the Mediterranean Fleet, but others contrived to find for themselves shore billets. Men from London Division thus found themselves in the Alexandria Provost Marshal's town patrol force, or assisting the Fleet Club Committee there, or acting as staff car chauffeurs, or crewing King's Harbour Master's launches. They had had a hard war and were lucky to come through the ultimate disaster: they deserved their respite.

London Division's gunnery link with the 'C' class A.A. cruisers, which had been forged in peace and tested in war, had ended with the death of the *Coventry* about half-way between Mersa Matruh and Alexandria. All four had acquitted themselves well, and all had died fighting. Forty years on it is gratifying to read in an authoritative book on British cruisers that the *Coventry* was 'one of the most effective A.A. ships in the entire British Navy, downing more aircraft than almost any other ship'[3].

Lieut. H. S. Whittow, who had led the London Division unit as it joined the *Coventry* in 1939, later served in the monitor *Roberts* as Gunnery Officer during the bombardment of Normandy in 1944. When, in the course of that duty, a shell jammed, his gun's crew managed to get it out and drop it over the side. At a subsequent Court of Inquiry when he was asked why he had not 'gauged' it with the ring-gauge he blandly replied, 'I had other things on my mind.' He was acquitted, and later awarded the D.S.C. In private life he was a Chartered Surveyor.

The Gunnery Officer of the *Cairo* when she was sunk during the 'Pedestal' Malta Convoy was Lieut. C. P. C. Noble, who had joined London Division as a Probationary Midshipman in 1932. Before he was mobilised in 1939 he had served during six successive Easter Cruises in the cruiser *Curacoa*, and therefore it was not surprising that at the outbreak of hostilities he commanded the London Division unit which went to her sister-ship *Calcutta*. He was awarded the D.S.C. after 'Pedestal'. Following a somewhat non-productive year as Lieut.-Cdr. (G) of the *Unicorn* in the Eastern Fleet he became the Staff Gunnery Officer, Western Approaches Training Flotilla, in the former Sopwith yacht *Philante*, which in 1945 took in the U-boats coming in from sea to surrender at Lock Eriboll and Scapa Flow.

A sad appendage to the story of the 'C' class cruisers concerns the death of Lieut. N. N. J. Bodger, who, as already recounted, went off with the FT unit of London Division to the *Cairo* just before mobilisation in 1939. At the end of 1940 he left the *Cairo* and went to do a Long Gunnery Course at Whale Island, after which he became the Gunnery Officer of the *Curacoa*. On 2 October 1942 she was escorting the liner *Queen Mary* off Bloody Foreland, Northern Ireland, and doing so from ahead on a steady course, with the *Queen Mary* zigzagging across her wake. Unaccountably the liner collided with her, rolled her over and cut her in half, with the loss of 338 lives, including Bodger's. A survivor said Bodger had just gone off watch and down to his cabin when the *Queen Mary* sliced through the *Curacoa*'s quarter deck, which, the survivor added, could be seen sinking a long way back. Many years afterwards the House of Lords held that blame was attributable to both ships, in the proportion of two-thirds against the Admiralty and one-third against the shipping line, the Cunard-White Star Company[4].

When E. D. Bennett joined London Division in 1934 as a Boy 1st Class he never dreamed that before he had done with the Navy he would be a Lieutenant-Commander, and R.N. at that; yet such was his personal saga. He was in the FT unit which went to the *Cairo* on mobilisation, but after a year he was sent to the *King Alfred* at Hove and received his commission. Thereafter his future lay with Coastal Forces. At Belfast in 1941 he was made First Lieutenant of a harbour defence launch nearing completion at Harland & Wolff's, but he soon got his own command – H.D.M.L. 1083 at Bideford. While patrolling in her in Rothesay Bay in December 1941 near the Asdic 'loops' guarding the entrance to the Clyde he received the signal 'Commence hostilities against Japan'. Later his motor launch was put on board a Lamport & Holt freighter at Liverpool in the charge of his First Lieutenant while he himself sailed for the Middle East in the *Empress of Japan* for the journey round the Cape of Good Hope.

In H.D.M.L. 1083 he patrolled off Alexandria, and whilst on the West Patrol from there on the night on which the Battle of El Alamein began (23 October 1942) he found himself only five miles to seaward of the action, and had an unforgettable view of the artillery bombardment with which it began.

On another occasion, an Asdic echo followed by a roar

RIGHT *Signal Cup Supper, 1908*

obtained by H.D.M.L. 1083 off Alexandria on a very dark night led immediately to great activity by Coastal Forces craft, and in due course to Bennett being interviewed first by the Commander-in-Chief (Admiral Sir Henry Harwood) and later by the Minister of State (Mr Casey), since it was believed that a U-boat might have been attempting to land a spy near the Montaza Palace to which King Farouk of Egypt had been removed from his palace at Ras-el-Tin. Bennett's superior officers were specially concerned that he had not dropped depth-charges at the time the echo was heard, because an inspection of the hull of H.D.M.L. 1083 revealed that a large area of barnacles had been wiped off her bows below water, presumably by the U-boat's jumping-stay as it dived hurriedly. Later, at Beirut, Bennett took over command of a 'Fairmile', H.M.M.L. 384, in which he served for two years, and finally as Divisional Leader of his flotilla of Coastal Forces in the German-occupied Dodecanese Islands.

Douglas Bennett's naval service was unusual in that, although he began it as an R.N.V.R. seaman, he was given an Extended Service Commission as a Lieutenant, R.N., in 1947, and obtained his 'half-stripe' as a Lieutenant-Commander, R.N., in 1950. He had gone to Java as a staff officer in 1946 and had served in the Far East in the mine-layer *Manxman* and the submarine depot ship *Mull of Kintyre* before becoming Assistant Queen's Harbour Master Hong Kong. He retired in 1953, and still resides in the City of London.

It was a fairly common occurrence for a pre-war trained rating of the Division to obtain a commission, but when it was gained it was only for the duration of hostilities, and 'Temporary' was placed before the R.N.V.R. rank held. The Division was in no way committed to retention of the holder on a permanent basis in time of peace, though it was possible to apply for such transfer after the war when the official opportunity to do so was provided.

Not all the pre-war ratings of the Division who obtained commissions did so in the earliest days of the war. Ordinary Signalman R. S. Noquet, who had joined the Division in February 1939, did not become a Temporary Acting Sub-Lieutenant until March 1943. After mobilisation he completed his signal training at H.M.S. *Royal Arthur*, the new-entry training establishment at the former Butlin's Holiday Camp at Skegness, and then served in two armed merchant cruisers, the *Corfu* and the *Maloja*, but after passing for Leading Signalman he transferred to Leading Writer and then went to H.M.S. *King Alfred* at Hove for his officer-training. He was in the *Bulolo* when she was H.Q. ship off 'Gold' Beach during the Normandy Landings, and then in the cruisers *Bellona* and *Diadem* on Russian convoys. He ended the war in Germany in a staff appointment, and had the strange distinction of serving for a time on board the German cruiser *Leipzig*, which was immobilised in dock after collision damage and was later towed out and sunk in 100 fathoms, in accordance with the Tripartite Agreement. When the Russians came on board to collect their share of the spares the shore-lighting failed. The Russians were immediately suspicious, and a nasty incident was averted when the Lieutenant, R.N.V.R., who was in command of the ship organised an impromptu party with large quantities of German gin.

The variety of tasks and experiences which befell the officers and men of London Division following mobilisation in 1939 is endless, and greatly exceeds the more limited opportunities available during 1914–1918. When Able Seaman E. R. Rogers, who joined the Division as an Ordinary Seaman in 1936, was mobilised he went with the QD unit to the *Calcutta* and experienced the events already described in which she took part. Then he joined the cruiser *Suffolk* just in time to share in the excitement of her 'enemy report' when she was the first British ship to sight the *Bismarck* in the Denmark Strait close to the Polar ice-edge. Later, in the same ship, he took part in the escort of convoys from Australia to Bombay, and spent some time with a working-party on a coral reef in Mauritius. In the destroyer *Carron* he laid mines off Norway.

On the way home from South Africa he stopped in Egypt, where he met with survivors from his old ship, the *Calcutta*. While there, he heard from one of them how Captain Lord Louis Mountbatten had administered justice to a London R.N.V.R. rating who, having gone to the *Kelly* from the *Calcutta* in 1940, had failed to call his relief when going off watch. The rating was Able Seaman J. D. Howarth, who had gained the D.S.M. off Norway. When the relief later came up as a defaulter and was asked by Mountbatten why he had failed to go on watch he explained that he had not been given a 'shake', whereupon Mountbatten asked Howarth why he had omitted to do this. 'It was really a skylark, Sir,' said Howarth. 'Oh, was it?' said Mountbatten. After a moment's reflection he said to the relief, 'You get seven days No.11' [extra work and drill]. Then he turned to Howarth and said, 'And you, Howarth, can do it for him'. One Sub-Lieutenant, three seamen and two signalmen of London Division had served in the *Kelly*, and all were saved when she went down off Crete.

When Rogers was demobilised he qualified as a Public Health Inspector. He then tried to join London Division as a Sick Berth Attendant (Health Inspector), but was told that was not possible. So he joined the Territorial Army and was promoted Staff Sergeant. Later he was commissioned, promoted Captain and placed in command of 340 Field Hygiene Platoon, R.A.M.C. (V), until he retired in 1970 having received the Territorial Decoration.

As is well-known, the (old) *Ark Royal* enshrined the honour of the Royal Navy during the first two years of the Second World War. She was regularly reported by the German radio to have been sunk in the various attacks which she suffered, but, as famous photographs show, she invariably came through the bomb-splashes intact, until the day when she was torpedoed, only twenty-five miles away from Gibraltar and safety. To her in August 1939 went two R.N.V.R. officers, and one of them was Lieut. J. G. Young of London Division. He was in the ship until the spring of 1941 and therefore was present at all her exploits up to that time.

The *Ark Royal* was attacked by the U-39 in September 1939, missed the *Altmark*, the *Graf Spee*'s supply ship, intercepted the *Uhenfels*, covered the landings in Norway and then the Narvik evacuation, operated against the French at Oran, covered numerous convoys to the eastern Mediterranean, was in action with the Italian Fleet off Cape Spartivento, operated against the French ships at Dakar, took part in the bombardment of Genoa, Leghorn and Spezia and the operation resulting in the sinking of the *Bismarck* – one of the *Ark Royal*'s aircraft hit her right aft and crippled her by wrecking her steering gear and jamming her rudders, thus sealing her fate.

Shortly after this, Young left the ship and went to Whale Island to do a Long Gunnery course, from which eventually he passed out top and was appointed to the staff there. He had no sooner settled in than he was given a 'pier-head jump' to Gibraltar to join the cruiser *Penelope* as her Gunnery Officer. This ship had established for herself a legendary reputation in the Mediterranean and her previous Gunnery Officer, Lieut. J. S. Miller, D.S.C. and bar, had been killed by the premature explosion of a shell in one of her guns during an air-raid on the docks at Valletta (his son is a member of London Division). The *Penelope* then went to New York for repairs, recommissioned at Portsmouth and after working up returned to the Mediterranean where she had made her name. There she took part in the landings in Sicily, at Salerno and at Anzio, and in the operations in the Aegean, where she was hit by a bomb which luckily failed to explode. She was finally sunk on 18 February 1944 by U-410 off Ischia, with serious loss of life, while on her way to support the Fifth Army, but Young was among the survivors.

Young returned to Whale Island, made a film on naval bombardment, trained hundreds of fighter pilots to spot for naval bombardment and was finally appointed Gunnery Officer of the 'County' class cruiser *Cumberland* in the East Indies. He had been one of the first R.N.V.R. officers to be appointed Gunnery Officer of a modern cruiser, and it was a fine R.N.V.R. achievement news of which, if it reached the envious Royal Naval Artillery Volunteers in Valhalla, was doubtless received with a standing ovation.

Commander C. B. Sanders, a tall officer known as 'Tiny' to

all his friends, had the distinction of commanding the big cruiser-minelayer *Adventure* for a period of six months ending in May 1944 (having previously been her Executive Officer), an achievement which was perhaps the nearest any R.N.V.R. officer came to commanding a large cruiser during the war. The *Adventure's* frequent task was to take 360 mines at a time from Milford Haven to the Mediterranean, where they were transferred in smaller loads to the fast minelayers, but she was later converted to a repair ship with 300 dockyard 'mateys' on board. When Sanders left her in August 1944 he went to the Admiralty, where, in the Department of the Director of Personal Services, he was attached to the Director of Manning and dealt with the drafting of ratings. Sanders became Executive Officer of the London Division soon after the war ended. He was promoted Captain, R.N.V.R. in 1949 and succeeded Captain R. T. Janson as Commanding Officer. He is a barrister who held several important offices within the City of London, becoming City Secondary (Under Sheriff), High Bailiff of Southwark and the General Administrator of the Central Criminal Court.

Another London Division member, Lieut.-Cdr. R. H. Green, who was a former Leading Seaman, became senior officer of a group of destroyers in the 'Hunt' class *Whaddon*, in which he was involved in operations and escort duties off the coast of Italy. In the First World War only one Volunteer officer commanded any naval ship bigger than a motor launch or a trawler, so these commands attained by Sanders and Green represented a big step forward for the R.N.V.R. and London Division[5]. The Executive Officer of London Division immediately before the war was Commander B. D. Nicholson. He became an Acting Captain and Station Officer at the Royal Naval Air Station at Donibristle during 1944, but in 1945 he was in Norway for the surrender of the German U-boats.

In addition to these examples of the attainments of executive officers during the war, some important appointments and fine work were achieved by specialist officers of London Division. To use again the traditional '*Navy List* Order' of 'parsons, doctors, pussers' which operated in the Royal Navy's more gracious days, examples of these are the Rev. G. E. Reindorp, who gained for himself a near-legendary reputation in the cruiser *Birmingham* in southern seas before 1942 and who went on after he retired from the R.N.V.R. in 1948 to be Bishop of Guildford and then of Salisbury; Surgeon Captain J. B. Ronaldson, who, having joined London Division as a Surgeon Probationer, R.N.V.R., in 1912, was a Surgeon Captain by 1930, was for much of the war Principal Medical Officer of the large New Entry Training Establishment *Royal Arthur* occupying Butlin's Holiday Camp at Skegness; Surgeon Commander G. F. Abercrombie, another former Surgeon Probationer from First World War days (when he was Mentioned in Despatches while serving in the destroyer *Warwick* during the attack on Zeebrugge), was also in the cruiser *Birmingham* and then in the battleship *Anson*; Surgeon Commander H. M. Willoughby, who in 1944 was promoted to Captain when he was appointed P.M.O. of the hospital ship *Oxfordshire*, and who, after the war, as Medical Officer of the Port of London Authority, also became P.M.O. of the Division; and Lieut.-Cdr. (S) F. A. Kemmis Betty, who, having survived the loss of the *Calcutta* on 1 June 1941, was then a survivor from the battleship *Barham* (in which 862 were lost) on 25 November, having been there in a temporary capacity. Later he was successively Secretary to Rear-Admiral (Destroyers) Mediterranean and Secretary to the Commodore-in-Charge, Sheerness, Rear-Admiral C. S. Thomson. After the war he became a Captain (S) and was Head of the *President*'s Supply and Secretariat Department. As all these officers had amassed great experience in their respective departments they held their own with ease among their Royal Navy colleagues, and they provide proof of the extent to which the Royal Navy had come to rely on the Royal Naval Volunteer Reserve by the time the war ended.

Approximately 17% (or one in six) of those mobilised from London Division during the Second World War gave their lives. When the reckoning came to be made the final total was 304. They are commemorated in a framed scroll by the entrance to the gangway of the *President*. It was dedicated at a Memorial Service held on 1 June 1949 in St. Martin-in-the-Fields, Trafalgar Square, in the presence of members of the Board of Admiralty, including the First Sea Lord, Admiral of the Fleet Lord Fraser of the North Cape. The Chaplain of the *President*[6] conducted the service.

London Division's Mobilisation Book listing every rating who was called up between 15 June and 6 December 1939 (on which date an Ordinary Seaman who, when first called, had been 'at sea in S.S. *Nebraska*' was drafted to R.N. Barracks, Chatham) was written up by the patient civilian secretary, Mr Reg. Knight, in a very clear hand. In the earliest months of the war he was careful to record any loss of life among the listed names. Some who died in the A.A. cruisers have already been mentioned, but others noted by Mr Knight's caring devotion before he put the book away may be allowed to provide a token, if woefully fractional, list of the Division's dead. He noted Leading Seaman P. Dorer, *Cairo*, carried away by a fast tide from a cutter alongside the ship in the Humber; Leading Seaman F. Pitkin, *Royal Oak*, lost when she was torpedoed in Scapa Flow; Signalman J. D. Bartlett, of the trawler *Kingston Cornelian*; and Signalman L. H. Stone, *Northern Rover*, another trawler, which was reported overdue at Kirkwall and later presumed lost. To these few names which were especially recorded we can add that of Sub-Lieut. P. Lancaster, one of the Division's first officer casualties, a trainee stockbroker, who was lost in the destroyer *Hunter* in the First Battle of Narvik on 10 April 1940, in the action in which his Captain (D), Captain B. A. Warburton-Lee, R.N., won the Victoria Cross.

CHAPTER SIX

Old Hands: New Navy

After the war formally ended on 2 September 1945 with the Japanese surrender in Tokyo Bay, a year was to elapse before the R.N.V.R. divisions were in a position to receive back their former members who wished to rejoin. The first priority in 1946 was Demobilisation, but after the Victory Parade had completed the wartime picture in June, Admiralty plans for reconstituting the R.N.V.R. divisions could be put into effect. On 13 July the Admiralty asked officers who held permanent commissions in the R.N.V.R. to write to their newly-appointed Commanding Officers. Many had already done this, perhaps with an eye to ensuring they had unbroken service when the time came to return on board the drill ships. The Admiralty's notification, which was widely published in the Press, made it clear that officers who had not indicated to the Admiral Commanding Reserves by 1 September their wish to serve in the post-war R.N.V.R. would be removed from the Active List. As a result, those who had signified their desire to do so rejoined early in 1947, while later in that year applications were considered within the divisions from officers who had held temporary R.N.V.R. commissions during the war and who wished to transfer to the permanent R.N.V.R. It was a matter of great pleasure for these hitherto 'temporaries' when they were informed at acceptance that they could retain their wartime seniorities. Many former temporary officers were thus able to join the R.N.V.R. on a permanent basis before 1947 was out, but as there were far more applicants than could be accommodated, particularly in London Division, not everyone who applied was accepted.

With London Division's officer establishment more or less settled, it was in a position in the autumn of 1947 to deal with recruiting, following the Admiralty's announcement on 25 August, which was also given wide Press publicity, that entry into the R.N.V.R. was now open to all who wished to volunteer, and was no longer confined to those who had previous wartime service with the Royal Navy. Few pre-war R.N.V.R. ratings had returned, and these were mainly petty officers, but they were valuable as a nucleus when it came to recruiting. National Service still continued, and opening the R.N.V.R. to all volunteers assisted young men due for call-up at the age of 18, for, provided they joined the R.N.V.R. at least twelve months before call-up, they were guaranteed entry into the Royal Navy for their National Service. At the same time direct entry into officer rank was discontinued, but volunteers who might expect to proceed to commissions could join as ratings in the branch of their choice. The age-limit for entry was twenty-six except for artificer and artisan grades, in which it was set at twenty-eight.

On 12 December 1947 the Admiralty announced its intention to establish a Volunteer Reserve for the Royal Marines, to be known as the Royal Marine Forces Volunteer Reserve, authority for which had been given by the Naval Forces Act of 1903. At about the same time as the R.M.F.V.R. was established, and again capitalising on highly qualified existing personnel, the Admiralty formed four R.N.V.R. Air Squadrons, but although one of these served the London–Oxford area it had no connection with London Division and was based at the Royal Naval Air Station at Culham, Oxfordshire.

During the war both the *President* and the *Chrysanthemum* had been used by the organisation known as D.E.M.S. (defensively equipped merchant ships) for the purposes of training Merchant Navy men and D.E.M.S. naval ratings, principally in the anti-aircraft weapons. For this purpose the *President* had been fitted with a grotesque-looking 'Dome-Teacher', a lofty and ingenious corrugated-iron structure, which combined a cine-camera for the instruction of A.A. gunners. A claim was made in the Press that this had trained between 70,000 and 80,000 men during the war years and was said to be one of the best wit-sharpeners in the Royal Navy.

Now that the *President* was to revert to her former purpose as London Division's main drill ship, she was sent for a big refit at Chatham on 20 July 1947. Commander J.G. Gould, R.N., the Division's Staff Officer, was in command of her during the journey along the Thames and Medway, and with him went her eight shipkeepers and a pilot. She was away from her accustomed mooring off the Embankment for almost a year, and, looking very different, she returned on 9 July 1948. Whereas at the end of the war she had become something of an eyesore, with many ugly wooden sheds on her upper deck, now she was transformed, with a new funnel, a stump mast in place of the two pole masts she formerly had, a lower deckhouse for the sake of symmetry, and a new bridge. Inside her she now possessed two 4-inch gun turrets with associated fire-control equipment, set well down in the bowels of the ship where the engine and boiler rooms had once been, and two Bofors guns. Plans had been made to fit her with a complete set of equipment for radar gunnery. The accommodation for training classes was greatly improved. The *President* had received a radical face-lift in appearance, and in fact was fitted out after the manner of a wartime destroyer.

The new Commanding Officer of London Division, Captain R. T. Janson, who had been appointed on 31 December 1945, was much relieved to receive back his drill ship, as was Commander C. B. Sanders, the new Executive Officer, who, as already related, had commanded the cruiser-minelayer *Adventure* in 1944. However, it slowly became apparent that the *President* had been refitted and re-equipped rather too soon after the war, on the assumption that training would go on in the manner customary in pre-war days, with emphasis on gunnery and communications. With National Service and a vastly reduced Fleet the concept proved wrong, and the new equipment was hardly used.

Between 1947 and 1950 R.N.V.R. training generally was at a low ebb and training in ships of the Fleet was hard to come by. As early as 1947 the Admiralty had allocated small coastal craft to divisions, with the intention that weekend training could be done in them, with longer periods during the summer months. Many officers had had wartime experience operating in destroyers, the motor torpedo boats of Coastal Forces and minesweeping trawlers and it was against this background of considerable expertise that the first tentative plans were laid for R.N.V.R. peacetime sea-training to move a step forward and take place in craft which the Reservists could man and operate themselves. Because of its size and the fact that not many of its officers were 'small ship' men, London Division was given the role of General Training.

London Division's first motor minesweeper, Thames. *(RN)*

(*News of the World*)

LONDON LAUGHS . . . *By* LEE

H.M.S. PRESIDENT

"Two to Richmond, please!"

London Division's first two sea tenders allocated in 1948 and 1949 and named *Isis* and *Thames* respectively (names which were to become traditional over the next few years for the Division's sea training craft), were fast despatch boats. One of these, the *Isis*, was replaced by the first motor minesweeper in the Division in 1950. She cruised to Dunkirk in May of that year under the command of Lieutenant H. E. J. Brown. A second motor minesweeper, replacing the despatch boat *Thames*, arrived in 1951, and possession of these craft added a new dimension to the training offered by the *President*, since Captain Sanders, who had succeeded Janson in 1949, was then able to devise annual training voyages to more distant and interesting ports. Great interest was taken in these years in amphibious exercises, the most ambitious being 'Red Dagger' when the Division planned and executed the landing of contingents from City Territorial Army units and the Royal Marine Forces Volunteer Reserve on the coast of Dorset with R.N.V.R. air support. A further exercise involved R.M.F.V.R. contingents embarked in the L.S.T. *Suvla* being landed in Stokes Bay in the Solent. The motor minesweepers *Thames* and *Isis* cruised extensively on training in the North Sea and beyond and were based on and normally moored alongside the *President* and *Chrysanthemum* respectively in King's Reach.

Of course there were also some opportunities of sea training being done with the Fleet in those early days, and in August 1950 the Chaplain of the Division[1], was appointed to the new battleship *Vanguard* for the period when she was guardship at Cowes Regatta, this proving to be the last occasion on which a battleship performed that duty, no doubt because her presence there occasioned a question in the House of Commons. In 1957 he was appointed to the aircraft carrier *Ark Royal* for her visit to the United States for the Naval Review which celebrated the 350th Anniversary of the Founding of Virginia in 1607.

The training of volunteers in peacetime to the standards necessary for command was an entirely new departure for the R.N.V.R. Although this decision was taken against a background in which the 'temporaries' of wartime days had shown it could be done, it was nevertheless imaginative and bold. So when in 1951 the Admiralty announced that R.N.V.R. officers were to be permitted to take the R.N. Modified Destroyer Command Examination, the Royal Navy ignited a spark which has truly motivated successive generations of volunteer seamen from all walks of life. In due course, a career structure evolved in which part-time volunteers could aspire not only to the command of their own ships, but also to fulfil the tasks of Squadron Commander, in overall command, on occasions, of eleven or twelve ships – something few full-time professionals could boast in peacetime.

On 30 March 1951 the unbelievable (and, some thought, the unforgivable) happened so far as the R.N.V.R. is concerned. An Admiralty announcement made on that day (the day of the R.N.V.R's Annual Ball at the Dorchester Hotel) said:

> The distinctive lace hitherto worn by the R.N.R. will disappear completely, and the R.N.V.R. wavy lace will be retained only for officers of the Sea Cadet Corps and the Combined Cadet Force. Except for this, all officers of the Reserves will in future wear lace similar to that worn by Regular Officers of the Royal Navy, but with variations in the curl . . . the traditional R.N.R. and R.N.V.R. buttons will be re-introduced at the same time.

Stating that as the modern Navy called for a wide variety of technical knowledge among both Regular and Reserve officers, the Admiralty said it had been decided it was no longer expedient, or in keeping with the times, to have any considerable distinction between the mark of rank of the Regular and Reserve officers. The 'variations in the curl' referred to were an 'R' or an 'A' (Air) worn inside the curl. The Admiralty also explained that all temporary officers would wear straight stripes but with a wavy curl – a hybrid which was a manifest absurdity, though not quite of the same order as the reintroduction of the R.N.V.R. buttons, which had been discontinued in 1922. A fortnight later the Admiralty announced that R.N.V.R. ratings would in future wear cap ribbons bearing the names of ships, instead of the then customary ones with the letters 'R.N.V.R.' followed by the names of their divisions.

When the order came to adopt the straight stripes the officers of the Division decided to have their old and now unwanted lace melted down and sold as scrap metal. They received £60 for it, and with the money they bought for use in the mess a silver cigarette-box, plus a snuff-box for the fastidious. As a spokesman for Messrs. Gieves, naval outfitters since Nelson's day, explained, the wavy lace contained 96% silver and only 2% gold.

The change of lace was alluded to in an amusing exchange of signals that occurred in March 1953 shortly after Captain J. A. Creed, an ex-Tyne Division officer, had assumed command of London Division. On 20 March the officers of the *Calliope* (Tyne Division) were holding their annual 'Samoa Dinner', which commemorates the escape of the *Calliope* from the disastrous hurricane at Samoa in 1889, unaware that on the same evening London Division were dining, for the first time, virtually the entire Board of Admiralty. During the evening Tyne dispatched the following signal:

> To: *President*
> From: *Calliope*
> On the occasion of the Samoa Dinner congratulations on attaining a Tyne Division officer in command.
> =202240

They were astonished to receive a reply, not from the *President* as expected, but from the Admiralty itself in the following terms:

> To: *Calliope*
> From: Admiralty
> Your 202240 to *President*. *President* is riding out an even heavier gale in entertaining the Board of Admiralty at dinner tonight. Isaiah Chapter 40 quote the crooked shall be made straight unquote.
> =202330

Throughout the evening the *President* proudly wore the Board of Admiralty flag at the mainmast.

Within a month of these uniform changes being announced, the motor minesweepers *Thames* and *Isis* were on their way to Lisbon on a training cruise, manned by six officers and twenty-four ratings from the Division. Captain Sanders, at 6ft. 7in. reputedly the tallest officer in the R.N.V.R., was with them, although he faced the ordeal of sleeping (when possible) in a 5ft. 10in. bunk for a period of fourteen days. Among the officers could be found a bank clerk, a stockbrokers' jobber, an inspector of taxes, a chartered surveyor and an Ealing doctor, all of whom were wearing their new straight stripes. It was a voyage of 2200 miles but the ships were only two hours late on their Estimated Time of Arrival. Three months later the two minesweepers went to Denmark and Sweden on another training cruise and with a different crew, which included a solicitor and a brewer, while fifty-three ratings from the Division joined the destroyer *Contest* for a fortnight's instructional cruise.

Before 1951 was out, however, the Division, in common with all other divisions, learnt it would soon have to 'stand by to receive boarders', and ladies at that; for the Women's Royal Naval Volunteer Reserve was founded in 1952. Discussions on its formation had taken place since 1949, the year in which the Women's Royal Naval Service was re-formed on a permanent peacetime basis. The Reserve was to model itself on its Service counterpart and as both were set up and administered initially by officers and senior ratings who had served together in the war

An early W.R.N.V.R. New Entry Class 1952. (RN)

this was not difficult. A nucleus unit was actually formed in London Division in November 1951, and Chief Officer Octavia Snow, who had taken part in the discussions setting up the W.R.N.V.R., was appointed to be the first Unit Officer – thus becoming responsible to the Commanding Officer for the discipline and welfare of the Wrens. Second Officer Joan Frame was appointed Recruiting Officer, and by June 1952 recruiting had to be temporarily suspended as the target of 100 recruits in the first year had already been reached. By 1954, numbers were at their peak with forty officers and over 250 ratings.

Although comparable with the W.R.N.S., the W.R.N.V.R. categories were fewer in number; in addition to Regulating and Dental Surgery Assistants they were mainly of the Communications and Supply and Secretariat branches. In 1957 a branch concerned with Degaussing, the science of demagnetising ships to protect them from activating magnetic mines, was introduced and allocated exclusively to the Reserve.

In addition to their training, which for the luckier ones could mean two weeks in Gibraltar or Malta, and more recently in Hong Kong, the W.R.N.V.R. took part in the sporting, ceremonial and social activities, and once a year were taken to sea for a W.R.N.V.R. weekend. Many a Wren will remember 'buffing up the brass' or 'peeling the spuds' on board a minesweeper on passage between ports on the South Coast, or en route to France or Alderney.

The advent of the W.R.N.V.R. had an electrifying effect on the Division, though many male diehards among the be-medalled officers and men had doubts as to how the scheme would work. The R.N.V.R. junior ratings had to vacate their changing space on the main deck forward of the *President* – a thing which would have caused their pre-war counterparts some astonishment. The idea of the invasion of the *President*'s monastic wardroom by Wren Officers was an unpopular prospect, but thanks to tact and good sense on both sides this feeling was short-lived, and the Division soon wondered how it had ever got on without Wrens.

In 1967 the Women's Royal Naval Service celebrated its Golden Jubilee with a historical pageant held at the Royal Festival Hall in the presence of H.R.H. The Princess Marina. London Division was asked to provide the cast, and rehearsals were held for several weeks under professional tuition. However, for some inexplicable reason, when rehearsing stopped at 6 o'clock on the evening of the performance, not only had the last Act not been rehearsed, but it had not even been cast. During supper, the senior officer present, script in hand, indiscriminately cast the various parts, and the Wrens, without hesitation, set about learning their lines. There was no chance of rehearsing, for the audience was already arriving. Stage directions were made and lines learnt in the wings seconds before appearance on stage. The fact that the pageant was a resounding success is a fine tribute to the W.R.N.R.

Apart from losing the 'V' and becoming the W.R.N.R. in 1958 there were no substantial changes until 1975, when, in line with other Divisions, the W.R.N.R. in London ceased to be run as a separate Unit and became an integral part of the R.N.R., the Wrens being administered by the departments to which they belonged. Apart from the sea-going role, the W.R.N.R. take part in all aspects of Divisional life, including former male preserves. The appointment of Administrative Officer, traditionally held by a Commander, was held by a Chief Officer in 1982. In 1977, the year in which the W.R.N.S. and the W.R.N.R. came under the Naval Discipline Act, the W.R.N.R. marked its Silver Jubilee with a Reception in Fishmongers Hall attended by past and present members of all W.R.N.R. Units. H.R.H. The Princess Anne had been invited, as Chief Commandant of the W.R.N.S., but found herself 'otherwise engaged' as it was held on 15 November, the day her son was born. At this time the total strength of the W.R.N.R. nationwide was 900 officers and ratings.

The Royal Naval Volunteer Reserve was fifty years old on 30 June 1953. Any Golden Jubilee celebrations then, however, would have fallen immediately after the Coronation, so they were postponed until the following year. Commander C. P. C. Noble, the Executive Officer of London Division, who had commanded the Reserves Marching Unit in the Coronation Procession of Her Majesty the Queen on 2 June 1953, was Parade Marshal on Horse Guards Parade for the R.N.V.R. Golden Jubilee Review on 12 June 1954, when Her Majesty the Queen, accompanied by H.R.H. The Duke of Edinburgh, inspected the 2000 officers, men and women who severally represented the R.N.V.R., the R.M.F.V.R. and the W.R.N.V.R. The Commanding Officer of London Division, Captain J. A. Creed, sat beside the driver of the Land Rover in which the Queen and the Duke stood resolutely as it passed up and down the ranks in heavy rain. That evening, at a memorable banquet held at the Royal Naval College at Greenwich, Captain the Earl of Iveagh, who had been the first Commanding Officer of London Division, proposed the toast to the Royal Navy to which the First Sea Lord, Admiral of the Fleet Sir Rhoderick McGrigor, replied. Lord Iveagh, then aged eighty, was the sole surviving officer who had joined the R.N.V.R. on the day it was formed, but he was accompanied on Horse Guards Parade in a place of honour by thirty-one officers and men from London Division who had joined during the R.N.V.R.'s first year of existence. Among these was Commander Harry Vandervell, who was born in 1870 and who belonged to the old Royal Naval Artillery Volunteers which were disbanded in 1892.

Less than two months before the Golden Jubilee Review,

The W.R.N.V.R. of London Division rehearsing at Duchess of Kent Barracks, Portsmouth for the R.N.V.R. Jubilee Review, 1954.

R.N.V.R. Jubilee Review on Horse Guards Parade 1954. The Captain of London Division, as Parade Commander, accompanied Her Majesty The Queen in the land rover.

H.M.M.S. 1558, borrowed by London Division for the weekend to replace its own motor minesweeper which had been withdrawn pending the arrival of the new coastal minesweeper, caught fire in the Channel off Ostend and sank at 0740 on 17 April 1954. The ship had sailed the previous evening (Good Friday) on a training cruise with a ship's company of 31 officers and men, the Commanding Officer being Lieutenant Commander P. W. T. Warren. A French steamer, the *Tunisie*, and the Dutch *Phoenix* picked up the survivors when they abandoned ship, fortunately without loss of life although one able seaman was taken to hospital.

Surgeon Commander (D) J. Symmons, later the senior dental

surgeon of the Division, who was on board as acting medical officer and mess secretary for the cruise, wrote an account of the disaster. He said that the weather deteriorated soon after H.M.M.S. 1558 sailed from the Thames, and the ship was leaking badly. A lot of water entered it, and the ship's safe broke away from the bulkhead to which it was fixed and moved about the wardroom with every roll of the ship. The galley cupboards broke open, and eggs, bread and meat spilled on to the wardroom deck in six inches of sea water. Having done what he could to rectify this situation Symmons smelt something burning. He went out on deck and saw flames ten feet high emerging from the ship's funnel. He made his way to the bridge and was told the engine-room was on fire and the ship was out of control. The ship was then between twenty-five and thirty miles from the nearest land and most of her crew were very wet and seasick. Gallant efforts were made to extinguish the engine-room blaze, but smoke and heat prevented success. As the ship had a wooden hull the fire was spreading, so signals for help were sent out.

When the *Tunisie* came on the scene she lay off 500 yards away and sent over a lifeboat. As the French crew were very seasick when they arrived Symmons was ordered to take command of it for the journey to the *Tunisie* in a force nine gale with an officer and thirteen men from the minesweeper, while the remainder stayed to fight the fire. Eventually the minesweeper was ablaze from stem to stern, and the order was given to abandon ship. Happily, all on board had been accounted for. The survivors were landed at Flushing and Antwerp and looked after by the Royal Netherlands and Royal Belgian Navies before returning to London via Harwich and Dover. The *President* was manned for the Easter week-end, so they were given a great welcome after their ordeal.

At a subsequent Court of Inquiry the Commanding Officer was absolved of all blame, for it was found that the condition of the minesweeper was such that it should not have been used for the cruise. It was thought that the sea water had caused electrical arcing. By all accounts the thirty-one officers and men who had been on board H.M.M.S. 1558 had had a lucky escape.

On the day of the Jubilee Review, 12 June 1954, the Admiralty announced that the Royal Research Ship *Discovery*, which had been Captain Scott's ship on his first voyage to the Antarctic in 1901–4, was to be taken over from the Boy Scouts' Association and used as an additional drill ship for London Division. The ship had been offered unconditionally as a gift, and the Admiralty reciprocated by allowing the Sea Scouts to keep their boats there, agreeing to provide them and other similar youth organisations with facilities on board during the forenoons on weekdays and throughout the week-ends. The Admiralty also stated that, while arrangements were being

Discovery at her berth in King's Reach wearing the flag of the Admiral Commanding Reserves at the foremast.

made to adapt the *Discovery* for naval service, it was intended to allow the public to have access to the ship. She went away to a graving dock at Blackwall for twelve months, and after she had been brought back to her old mooring alongside the Victoria Embankment not far from Waterloo Bridge she was commissioned as a naval vessel on 20 July 1955.

On commissioning, the *Discovery* wore the flag of Admiral Commanding Reserves, and was used by the Division chiefly for those members who were doing their pre-National Service training. When National Service was abolished in 1960 the *Discovery*'s role in the Division's programme became much reduced, although she continued to fly A.C.R.'s flag until shortly before that post was abolished on 1 January 1977. Not long after she was taken into naval service certain Heads of Departments in the Division were given the use of the wooden cabins which are grouped round her wardroom and which bear the names of leading members of Scott's Expedition. Perhaps not all who used the *Discovery* at that time were aware that the cabins were not in fact those which were in the ship in Scott's day. The original historic cabins were removed in the early Twenties when the ship was converted for use as a store-ship for the Hudson's Bay Company, the old wardroom becoming a hold. Her original charthouse also did not survive. Before the Division finally gave up the use of the *Discovery* at a formal ceremony on 2 April 1979, when she was handed over to the Maritime Trust for preservation, the ship had been used as the naval recruiting centre in London (for which Admiral Commanding Reserves was Director for some years) for Reserve new-entry training, and from time to time as a substitute for the *President* when the latter was under refit.

In view of the recent naval activity in the South Atlantic perhaps it should be recorded that from 1925 the *Discovery* belonged to the Government of the Falkland Islands and was a Royal Research Ship. In 1937 she was presented to the Boy Scouts' Association by the Colonial Office and the Legislative Council of the Falkland Islands, as a memorial to Captain Scott and other Polar explorers.

The precise role the R.N.V.R. should have in the post-war Navy was established in 1954, when the far-reaching decision was made to allocate to R.N.V.R. Divisions the then brand new 'Ton' class coastal minesweepers, which had only just come into service and were to prove one of the most versatile and long-lasting classes of post-war years. London Division was allocated

Permanent Staff Instructors of London Division lunching in Discovery*'s wardroom shortly before she was formally handed over by the Navy to the Maritime Trust for preservation in 1979.*

Her Majesty The Queen, accompanied by The Duke of Edinburgh, Prince Charles and Princess Anne, passing the drill ships at the end of her Commonwealth tour in 1954.

Alverton, which was renamed *Thames* on 8 June 1954. The arrival of these vessels meant that sea-training throughout the R.N.V.R. could be rationalised and co-ordinated towards making a single squadron of operational ships, then twelve in number (one for each division), manned entirely by Reservists. This Reserve Squadron was to stand alongside its Royal Navy counterparts fully trained, and, together with them, was later to be committed in time of emergency or need to N.A.T.O. It is a concept unique to the Royal Navy, which has stood the test of time, and today, very nearly thirty years on, the 10th Mine-countermeasures Squadron, still manned entirely by Reservists, represents more than half the nation's capacity to meet the threat of mines in the approaches to its ports.

This policy gave a sound political and economic justification for the continued existence of the R.N.V.R., since the service it provides costs the nation a fraction of what would be needed to provide it full time. It also gave Reservists themselves a real and identifiable role in the naval defence of the Realm. This was a challenging and thoroughly healthy state of affairs for an organisation which depends entirely on voluntary commitment with few sanctions to apply if that is not met.

These far-reaching decisions, however, had barely been implemented when, in 1956, on the eve of the Suez emergency, economic stringency in Defence spending was demanded by the Treasury. Admiral Commanding Reserves, then Rear Admiral G. Thistleton-Smith, was asked to save a million pounds a year from the R.N.V.R. budget. This was a serious moment for the Reserve, for although the long term plans for its future role were

The Division's first coastal minesweeper, Thames *(ex* Alverton*).*

sound, as was later to be proved, these had not yet been implemented for long enough to show that this was necessarily going to be the case. The savings demanded led in 1958 to the closing down of Humber Division and the merging of the R.N.V.R. with the (old) R.N.R., which latterly had existed solely for officers of the Merchant Service. The formation of the new combined Royal Naval Reserve, with effect from 1 November 1958, was based on an idea expressed in a paper written some years earlier by Captain H. G. Boys-Smith, R.N.R. and Captain Sanders, R.N.V.R. shortly after the latter had relinquished command of London Division. Both sides of the Reserve treated the reform with caution. The professional merchant seamen objected to being linked with the 'enthusiastic amateurs' and the 'enthusiastic amateurs' were loath to lose their distinctive and historic style of 'volunteer'. But in truth both were 'volunteers', whether amateur or professional seamen, in their devotion to the Royal Navy, and in the years that have passed since the merger was made the bond of understanding and respect between the two has steadily grown stronger.

Also in 1956 the motor minesweeper *Isis* was replaced by the 'Ham' class inshore minesweeper *Pulham*, which was renamed *Isis*. In the summer of that year, under the command of Lieut.-Commander R. H. Bradley, she bravely steamed across the North Sea from alongside the *President*, and then proceeded 280 miles up the Rhine on a 'show the flag' visit to the Royal Marines at Krefeld. This was a demanding fourteen days, for the Rhine is a swift-flowing river with many treacherous currents. It was so successful, however, that it was repeated the following year with a different crew.

The Sixties saw the development and fulfilment of the policies laid down in the mid-Fifties. The 10th Minesweeping Squadron regularly took part in N.A.T.O. and Home Fleet exercises and the pattern developed of fourteen-day training periods which were divided into two parts. The first week was taken up by the exercise, during which the crews worked extremely hard, and at a convenient moment in the second week a foreign visit was arranged to lighten the burden and encourage the volunteers to 'come again next year'. Sea-training was largely allocated to the summer months, not because the Reservists were in any sense 'fair weather sailors' but because the ships they operated were small and, in order to make the most of training opportunities, they needed good weather conditions. Ships were traditionally refitted during the winter months.

The standards achieved were such that successive Admirals Commanding Reserves could propose the Reserve ships for operational tasks confident that they could, and would, perform as well as their Royal Navy counterparts. Thus in the summer of 1962, when a cruiser could not be spared for an official 'goodwill' visit to Bayonne, a squadron of six Reserve minesweepers was despatched under the command of Captain I. A. B. Quarrie, then Captain of London Division, to take her place. In 1963 when the Royal Navy was given the duty of assisting in clearing areas of the North Sea, which had not been swept for mines since the end of the war, to enable an underwater telephone cable to be laid, the 10th Minesweeping Squadron was one of the three British squadrons allocated for this service. For the *Thames*, commanded by Lieutenant Commander B. J. West, and the R.N.R. as a whole, it was the first time since 1947 that Reservists had been asked to tackle 'live' mines as part of their annual training. Captain B. J. Anderson, R.N., in command of the British squadrons aboard the support ship *Reclaim*, is reported to have said at the time 'I found no difference at all between the Royal Navy and the Reserve ships' – a fine tribute to the level of training and professional skill achieved by the part-time volunteers.

In 1964 *Isis* provided a naval presence on the occasion of the Queen's Birthday celebrations at the British Embassy in Paris, attended by the Duke and Duchess of Windsor and Earl Mountbatten. The voyage started from alongside the drill ships in King's Reach and finished in the River Seine almost directly alongside the Eiffel Tower. Afterwards Sir Pierson Dixon, the Ambassador, wrote: 'The smart turnout and bearing of the men aroused the admiration of our guests, and did much to enhance the ceremonial side of the occasion.'

On the occasion of the funeral of Sir Winston Churchill on 30 January 1965 the officers and men of London Division were well placed to pay their homage, as the coffin of the man who had been First Lord of the Admiralty during two world wars

Isis *under the command of Lieut.-Commander R. Richardson leaving King's Reach at the start of a visit to Paris in 1965. Note the 'jury rig' mast to enable her to navigate under bridges.*

ABOVE *With bowed heads London Division pays final homage to the 'Former Naval Person' as his coffin passes the* President *on its way from St Paul's via Waterloo to Bladon. (A.P.)*

London Division marching with arms reversed in the funeral procession of Sir Winston Churchill on 30 January 1965.

passed up river from Tower Pier to Festival Pier near Waterloo Station. Having provided earlier in the day an armed escort in the funeral procession consisting of six officers, six senior rates and twenty-four junior rates, they manned the side in both the *President* and *Chrysanthemum* as the Port of London Authority launch *Havengore* passed on its sad journey. Over fifty years previously at the drill hall in Commercial Road, Lambeth, he had spoken with much enthusiasm about the 'quality, intelligence and efficiency' of the officers and men of London Division, and now the successors of those men were returning the compliment in silent respect.

It was a busy and exciting year. Owing to the crisis in the Far East, the Royal Navy found itself short of a frigate to maintain the Caribbean/West Indies Station. The assistance of the Reserves was sought, and, under the command of Commodore Sir John Clerk, Bt., and Captain J. B. Leworthy (then Captain of London Division), four coastal minesweepers of the Reserve Squadron sailed on an eight-week trip across the Atlantic to the West Indies and British Guiana, maintaining a naval presence where needed until relieved. So keen were Reservists to go, perhaps not unnaturally, billets were substantially oversubscribed and many of the sailors that where chosen gave up their jobs to be able to join. The *Thames* did not take part in this transatlantic adventure, but later in the year, with other ships of the Squadron, she attended the Queen's Review of the Fleet in

Exercise 'Rockhaul' – The Reserve Squadron entering Gibraltar Harbour. *(RN)*

the Clyde, and went on to visit Oporto in Portugal. On the voyage home, whilst proceeding up channel in freshening weather with the *Curzon* and the *Warsash*, as darkness fell, she rescued a shark-fishing vessel adrift off the Lizard with an engine breakdown, taking off a party of ten holiday-makers which included women and small children.

The early Sixties also saw the start of the now legendary 'Rockhaul' and 'Mainhaul' exercises to Gibraltar and North Africa. The first one took place in the summer of 1963 when the Admiral Commanding Reserves, Rear Admiral H. C. Martell, 'borrowed' the cruiser *Belfast*, manned her with Sea Cadets, and sailed for the Mediterranean with his eleven Reserve minesweepers in company, including the *Thames* under the command of Commander A. H. Spratt-Kerswill. This imaginative fourteen-day training period concentrated on general seamanship drills and evolutions and was a major attraction for new entries. It was supported by a base-party made up of Reservists, including Wrens, who flew out to Gibraltar. In later years these exercises developed into a four-week deployment with a changeover of crews (doing fourteen days each) at Gibraltar.

This concept of utilising the more settled Mediterranean weather for training paid handsome dividends in providing both recruits for the R.N.R. and ideal working conditions. Furthermore, having the ships in the Mediterranean presented new opportunities for their use by the Royal Navy on N.A.T.O. exercises. In 1970 an extra two weeks was tacked on to the training period of the second crew. There was no shortage of volunteers to serve for a month. This enabled the Royal Navy, which had no other ships available at the time, to use six ships of the Reserve Squadron, including the *Thames*, to fulfil the United Kingdom commitment to an exercise involving units of the

W.R.N.R. enjoying a relaxed moment in Gibraltar. *(RN)*

A 'run ashore' in North Africa. (RN)

American and Italian Navies off the coast of Italy. Their performance was such that, at the debriefing on completion of the exercise, as the Commanding Officer of the *Thames* recalls, the Italians and Americans resolutely refused to believe that the British ships were manned entirely by Reservists, with no regular personnel on board except perhaps a chef or two. They thought it was some wicked British joke!

It is interesting to reflect that in the years before the Second World War the senior officers of the Division regarded the offices they held as 'freeholds' until they reached retirement age. Captain King commanded throughout the Twenties, and continued to do so after he became a Commodore, and Captain

The Reserve Squadron under way. (RN)

Hemsted commanded from 1933 to 1939, only ceasing then because the war had intervened. However, after the war, by a 'gentleman's agreement', Commanding Officers of the Division limited their time in command to three years, and this has now become the established norm. Again, before the war there were only two lists for officers – List 1 and List 2, the former being active – but by 1970 bureaucracy had devised as many as fourteen. Today there are only five, List 1 being the training list of Merchant Navy officers and Lists 3 and 4 the principal training lists of the civilian volunteers.

Formal occasions held at midday are necessarily rarities for London Division, but the one which was held on 28 April 1970 was rare indeed. On that day the senior officers of the Division received on board the then Admiral Commanding Reserves and Director of Naval Recruiting, Rear-Admiral B. C. G. Place, VC, who on behalf of the Division accepted from the Tisdall family the Victoria Cross won by Sub-Lieutenant A. W. St. C. Tisdall, R.N.V.R., during the landing from the *River Clyde* at 'V' Beach at Gallipoli on 25 April 1915. Many members of the Division who served during the First World War, and especially in the Royal Naval Division to which Tisdall belonged, were present to witness the presentation of the medal, which is on permanent loan to the Division and currently on exhibition at the Naval Museum in Portsmouth[2].[3]

The Tisdall Exhibition mounted at the gangway of HMS President.

CHAPTER SEVEN

Stop-Go Years

There can be little doubt that the Sixties and early Seventies were marvellous years in which to be a sea-going volunteer. The training voyages, which for London Division were four and sometimes five per year, each of fourteen days duration, were purposeful and often physically exhausting. Apart from the Mediterranean highlights mentioned in the last chapter they included annually a highly concentrated period operating from HMS *Vernon*, the Royal Navy minesweeping school at Portsmouth, and one, sometimes two, N.A.T.O. exercises operating out of Plymouth, the Forth or a continental port.

The Reserve (10th Minecountermeasures) Squadron had achieved a great deal since its inception fifteen years earlier. Those in the Royal Navy, who had worked with it were continually impressed by the young volunteers' enthusiasm and eagerness to learn. It was a buoyant and confident London Division that, on 5 May 1972, celebrated its twenty-five years of achievement since re-formation in 1947 with a dinner attended by the Lord Mayor, Sir Edward Howard. In his speech that evening the Lord Mayor said:

> Of the three Services, the Navy is closest to the City. This is a long tradition, for the Navy Office was in Seething Lane when that very City gentleman, Samuel Pepys, became Secretary to the Navy and used to go down from there to Deptford . . . Prosperity depended on trade, trade depended on ships and ships depended on sailors. In turn our ships and our sailors depended on the Royal Navy to ensure the freedom of the seas.
>
> It is no accident, therefore, that as Lord Mayor I should bear the honorary title of Admiral of the Port – the only Admiral in the Royal Navy with a moustache!
>
> The spirit of 'service' is exemplified by you. We in the City are very conscious of the work you do and are very proud of the service you render.

FAR LEFT *Wrens afloat under training in Kings Reach.* *(RN)*

THESE PICTURES AND FOLLOWING PAGE *Reservists undergoing sea training.* *(RN)*

25th Anniversary Dinner – on board HMS President. *From left to right: Captain West; The Lord Mayor, Sir Edward Howard; Rear Admiral McLaughlan, Admiral Commanding Reserves; Commodore Young; Captain Leworthy; Captain Creed; and Captain Quarrie.* *(City Press)*

The early Seventies marked the high point in the fortunes of London Division, and indeed of the whole R.N.R., since the end of the Second World War. With a few exceptions the Commanding Officers of the *Thames* and the other ships of the Reserve Squadron, who with their crews had achieved so much, had either seen some full-time service in the Royal Navy in the Second World War or had spent at least two years in the Royal Navy during National Service. Looking ahead into the Seventies and beyond, when this expertise would no longer be available, it was less clear what the future held. Furthermore the 'Ton' class coastal minesweepers in many instances were coming to the end of their useful life. Equipment was requiring more and more attention, as it, too, was old and liable to failure, thus reducing training value and morale. A new generation of ship to replace the 'Ton' class coastal minesweepers had not yet been decided upon, and fresh doubts were openly expressed as to whether the next generation of volunteers, who had seen no full-time service at all, would be able to master the technology involved in commanding and operating new ships and new gear.

Just as twenty years earlier, in the mid-Fifties, the sea-going role of the R.N.V.R. had been imaginatively defined against a background of some who doubted the capacity of the Reserve to cope, so again in 1974 fresh voices were to be heard asking where the R.N.R. was going, and what could it realistically do in future to support, particularly at sea, the operational tasks of the Royal Navy. It was undoubtedly time that the role of the Reserves should be thoroughly reviewed, and at the instigation of the then Admiral Commanding Reserves, Rear-Admiral I. G. W. Robertson, the Royal Naval Review Committee came into being on 1 March 1974. This Committee, which became known as the Mitchell Committee after the name of its Chairman, Rear-Admiral G. C. Mitchell, consisted of six members, a Captain and Commander Royal Navy, two civil servants and two Royal Naval Reserve Officers, Commander F. D. Patterson from Tyne Division and Captain A. B. Sainsbury originally

LEFT *HRH Prince Philip arriving to dine with the Chief and Petty Officers in 1970.* *(RN)*

BELOW *Admiral of the Fleet Lord Mountbatten addressing the Ship's Company after presenting the Mountbatten Trophy to the Communications Department in 1972.* *(RN)*

from Tyne Division but subsequently also from London Division where he was an ex-Head of the Supply Branch.

The Committee's task was to identify the role of the R.N.R. in peace and war over the next twenty years. In that context it had to recommend both the composition and structure of a re-organised Reserve and how this should relate to the Royal Navy, which, of course, meant that, among other things, the post of Admiral Commanding Reserves itself was under review. The Committee took nine months to reach its conclusions, having consulted widely amongst Reservists and others in the Royal Navy. Eventually there was unanimity in the recommendations that there must be a sea-going role for the R.N.R., that there must be an element of 'fun' (a word that took a lot of acceptance) in the training of volunteers, and, above all, that the R.N.R. was not a private navy but 'the part-time element in a single naval service'. To further this latter concept the Committee saw no need for an independent Flag Officer (Admiral Commanding Reserves), reporting directly to the Admiralty Board via the Second Sea Lord. It therefore recommended that the Royal Naval Reserve should be absorbed directly into the central naval organisation under the Commander-in-Chief Naval Home Command in Portsmouth, with certain delegated responsibilities going to Area Flag Officers. The operation of the sea tenders was to pass directly to the Commander-in-Chief Fleet at Northwood, who was responsible for all operational naval vessels afloat. These recommendations were ultimately accepted by the Admiralty Board and widely welcomed by the Reserve, but two years were to pass before implementation was complete. During this time London Division was to suffer severe setbacks.

The first of these occurred on 9 May 1974 when the Captain of the Division, Captain R. St. J. Fancourt, who was the first Master Mariner to command the Division, was knocked down and seriously injured whilst crossing the Embankment opposite the drill ships. Fortunately he ultimately made a complete recovery, but the accident resulted in his absence from the Division at a critical time. The executive officer, Commander A. P. M. Woodward, was immediately appointed Commander-in-Command and undertook the many additional responsibilities involved at very short notice.

The second setback affected the whole R.N.R., for whilst the Mitchell Committee was deliberating, and before it had reached a conclusion, the Government of the day announced a drastic reduction in Defence expenditure. The Reserve Forces had, of course, to accept their fair share of the economies demanded by the Treasury, and when they came in 1975 they cut deeply into the fabric of the R.N.R. The Reserve Squadron was gradually reduced from eleven ships to six. In addition to the loss of ships, in the summer of 1975 it was announced that all Divisions were to be reduced in complement. Captain Fancourt, now back in harness, was faced with the unenviable and almost impossible task of deciding who was to be allowed to remain and who was to go. The Technical and Supply Departments were the most seriously affected, and many worthy volunteers of long standing who could not be found billets in other branches had to leave (despite the fact that to keep them on a non-training list would have cost the Crown nothing). The Admiralty even considered scrapping the *President* at this time, which was then approaching its sixtieth year of life, but it was found that to do so and to convert the *Chrysanthemum* to do the work of both ships would cost a million pounds, and so the plan was eventually shelved.

The 'Reserve' (10th Minecountermeasures) Squadron formed into 'Groups' to remain viable, Divisions having to share vessels in order to maintain sea-training capability. (The justification given for this by the exponents of economy was greater utilisation of fewer ships.) London Division became part of the Channel Group which consisted additionally of Solent and Sussex Divisions. As the number of ships diminished, this necessarily meant fewer training opportunities, and the men no longer knew their own ship as they used to. London Division, with the greatest training requirements in terms of numbers of any Division in the R.N.R., was amongst the first to lose its sea tender. The last training period undertaken in the *Thames* was that of 7–9 November 1975, which incorporated a visit to Jersey to take part in local Remembrance Sunday ceremonies. Four years were to pass before London was again to have a dedicated

The Captain's day cabin in HMS Chrysanthemum.

The last HMS Thames *withdrawn from service in November 1975.* *(RN)*

coastal minesweeper. Although sea-training continued it became fragmented and irregular and the expertise built up over the past years, which depended so much on teamwork, *esprit de corps* and the sailors being able to take a pride in their own ship, began to slip away.

Such ships as remained allocated to R.N.R. divisions, albeit on a group-sharing basis, reverted back to their original Royal Navy names. This meant a break in tradition, for, from now on, any ship allocated to London could no longer be re-named *Thames* (or *Isis*) as had occurred regularly since 1948.

It was against this general background, although it had no direct bearing on what happened, that London Division suffered yet another blow when the worst naval disaster for twenty-five years occurred in the North Sea on 20 September 1976.

In mid-afternoon of that day the coastal minesweeper *Fittleton* (ex-*Curzon*) (360 tons) maintained by Sussex Division but manned at the time mainly but not exclusively by London Division, collided with the frigate *Mermaid* (2,300 tons) during manoeuvres some 80 miles north of Texel off the Dutch coast. The *Fittleton*, together with six other ships of the 10th Mine-countermeasures Squadron, had completed her part in the N.A.T.O. exercise 'Teamwork', which had involved some 275 ships, and was en route to Hamburg on a scheduled foreign visit. The *Mermaid*, under the command of Commander R. H. C. Heptinstall, R.N., had just joined the formation wearing the flag of Admiral Commanding Reserves, Rear-Admiral H. W. E. Hollins, and all ships were waiting with keen anticipation for their Admiral's first orders. The *Mermaid* took up a position two and a half cables to starboard of the seven minesweepers, which were in column, abreast the centre ship. All ships were proceeding on the same course at a speed of eleven knots in reasonably calm weather, though there was some swell. When a 'heaving line' transfer was ordered there was delay over interpretation of the signal and when the Admiral asked 'What are you waiting for?' the *Fittleton*, the first ship in the column, turned to starboard to execute the 'Z' type manoeuvre necessary to fall back towards the *Mermaid* and thereafter make for the replenishment point just aft of the *Mermaid*'s port bridge wing. This point was unusually far forward on a frigate of the *Mermaid*'s class, for, although her hull was standard, she had originally been constructed for the President of Ghana, before being modified for use by the Royal Navy, and her superstructure was therefore unique. As the *Fittleton* steadied on a course parallel to the *Mermaid* and began to edge-in close enough for a heaving line to be thrown, 'interaction' between the hulls of the two moving ships took charge, and sucked the hulls together in a minor side-to-side collision.

It was not until after this that disaster struck, but when it did it was swift and complete. The Commanding Officer of the *Fittleton*, finding his ship edging forward on the *Mermaid* despite efforts to stop it, decided to try and break free by going ahead and to port. As the *Fittleton* drew clear ahead of the *Mermaid*, however, she suddenly without warning slewed to starboard across the latter's bow. Commander Heptinstall immediately rang full astern, in an effort to reduce speed and avoid contact, but there was little he could do. The *Mermaid*'s bow struck the starboard side of the *Fittleton*, aft, and rolled her over.

It was inevitable that as the ship rolled over and settled upside down many would be trapped below deck. It says a great deal for the discipline and courage of the forty-four men on board that no less than thirty-two were saved. Some, of course, were thrown into the water from the upper deck, including the Commanding Officer and his bridge team, as the *Fittleton* was an 'open bridge' ship, but others were below in the lower mess decks. With unbelievable coolness they reorientated themselves to what had happened, opened watertight doors – taking care to close and clip them afterwards – and made their way down into the water towards the upper deck to get out and swim clear. Miraculously the emergency lighting remained working, greatly assisting those groping their way down towards the exits under water.

The *Mermaid* stopped very quickly indeed, thus ensuring no one in the water was injured by her propellers. A number of minesweepers closed the upturned hull of the *Fittleton* to assist.

The nearest was the *Crofton*, under the command of Lt. Commander M. Howard Tripp of Solent Division. In the words of one of the survivors at the time, '*Crofton* was fantastic . . . She shut down her engines and drifted in. If she had not acted so quickly a lot more people would have died.' Sadly, twelve men did die, including three who were trapped in the machinery-space unable to get out of the upturned hull. Although the hull remained afloat for some hours no one at the scene knew for how long the hull would stay buoyant, for it could have foundered at any moment, and very difficult decisions had to be made by those in charge of rescue operations. There were acts of heroism where men put their own lives at risk in order to help others to escape. CMEM Bissington of London Division ultimately received the Commander-in-Chief's Commendation for his attempts, shortly after reaching the surface himself, to dive down and open the machinery control room hatch.

At home news of the accident broke first on the TV news programmes of that same evening. By the following morning it was front page banner headlines in all national newspapers. Early that morning, in an expression of deep concern, the Second Sea Lord, Admiral Sir David Williams, came down in person to the *Chrysanthemum* to call upon Captain Fancourt, and he met the executive team, made up of permanent staff and Reservists, which had been up all night manning telephones, answering queries and visiting bereaved families. He offered any help the Division required and instantly made available personnel from his own staff to assist.

The *Fittleton* had been commanded by Lieut.-Commander P. J. V. Paget of London Division. He held a Merchant Navy Master's Certificate of competence and had been in the Division for sixteen years. In this time he had gained considerable experience in the handling of minesweepers. Within a short time of the incident a Royal Naval Board of Inquiry was held in Portsmouth at which all the survivors gave evidence. This Inquiry was behind closed doors. It was clear, however, that a public investigation of all the facts leading to the tragedy had to be mounted. The traditional Naval way of doing this is to court martial the Commanding Officer. It was a strongly held view both in the Division and outside that the Reserves should be treated no differently from the permanent service when it came to disciplinary matters of this kind. It was with some relief, therefore, that the Division heard of the decision to court martial Lieut.-Commander Paget – not least by Paget himself who was anxious that the whole matter should be investigated fairly and fully in open court.

On 20 February 1977 Paget appeared before a Court Martial held in Barham Block, H.M.S. *Nelson* in Portsmouth, on four charges of hazarding his ship. Proceedings lasted five days, and late on the evening of the final day, Paget was found guilty on one charge but cleared on the remaining three charges, which included the most serious one of allowing his ship to be lost. He received a reprimand which was the lightest sentence possible. However, on 11 May 1977 the Admiralty Board of the Defence Council, after reviewing the proceedings, pronounced even this conviction 'unsafe and unsatisfactory', and quashed it. The Board also annulled the sentence, saying that although there had been errors of judgement it was satisfied these did not amount to negligence under the Naval Discipline Act. The fact that Lieut.-Commander Paget was finally exonerated emphasized the problems inherent in manoeuvres of this kind, and so far as 'interaction' was concerned, confirmed that there were lessons to be learned from the incident by all who operate small vessels in close proximity to larger ones whilst under way. Thoughout this terrifying ordeal Paget had behaved with exemplary courage. His main concern was always for the welfare of the families of those who lost their lives. The tragedy had the effect of binding the whole Division together, and everyone set about thinking of ways to help.

On 28 September 1976, only eight days after the tragedy, Commander (later Captain) P. S. Rees became Commanding Officer of London Division. He was the first Instructor Officer to achieve command in the R.N.R. and the first non-seaman officer to take over London Division. Before Captain Fancourt relinquished command, he set up the H.M.S. *Fittleton* Fund for the benefit of the bereaved families. The Fishmongers Company and the City were among the first to contribute, and their contributions were quickly augmented by sums large and small from well-wishers all over the country and abroad. So far as the Division itself was concerned the main effort went into two events, both of which were supported by many individuals and organisations outside the Division. A sponsored bicycle ride from the *President* to the village of Fittleton in Wiltshire, a distance of 93 miles, took place on 5 February 1977; and the 'Fittleton Fayre' was held in Duke of York's Barracks, Chelsea, on Saturday, 26 March. This was opened by the actress, Susan Hampshire, and attended by other stars of stage and screen, including Lieut.-Commander Richard Baker, R.N.R. When the Fund was closed it had reached over £80,000, all of which was distributed to the families concerned. A Memorial Service for those who were lost in the *Fittleton* was held in St. Martin's-in-the-Fields on 30 November 1976, and it was attended by the Commander-in-Chief, Naval Home Command, Admiral Sir Terence Lewin, and the then Lord Mayor, Sir Robin Gillett.

For much of 1977 there were few opportunities for London Reservists to get to sea, though by then the Reserve Squadron had grown to seven coastal minesweepers and two 'Bird' class patrol vessels. London Division commissioned the patrol vessel *Sandpiper* into service at a ceremony held in St. Katherine's Dock on 16 September 1977. However, the training role of these vessels was limited as they were a Royal Air Force design, intended basically for day running.

To maintain morale the Division turned to other things. Training in non-seagoing areas carried on as normal, and throughout the late Seventies the growth continued of the N.C.S. Branch which had been established in January 1973 specialising in Naval Control of Merchant Shipping in time of emergency, certain lucky members of which have been known to travel as far afield as Hawaii in the Pacific for their annual fortnight. The medical and dental specialists who could boast skills such as underwater and aviation medicine continued to be available for recall each year to assist the permanent service during leave periods and when circumstances required additional medical back up. Medical support for the City of London Royal Marines Reserve is provided by London Division and the doctors found themselves on many assignments, from Northern Norway to Belize in British Honduras. It was on an Army expedition in Northern Canada in August 1972 that Surgeon Lieut.-Commander P. N. Dilly of London Division was responsible for saving the life of a fellow expedition member trapped in a crevasse some 160 feet deep, for which he was ultimately awarded the George Medal.

Communications officers and ratings continued to train on annual exercises and provide back up for the manning of Port Headquarters and Communication centres with members of the Royal Naval Auxiliary Service. Specialists in Security and Intelligence quietly went about their business – all, whatever their specialisation, being part-time volunteers with civilian jobs to do.

During the lean years in the late Seventies, the Royal Navy tried hard to maintain the operational viability of the Reserve Squadron. Two trawlers were chartered for use by Cardiff and Bristol Divisions in developing deep water minesweeping techniques. Minehunting, a special skill involving sensitive sonar equipment, was introduced as a further area in which naval reserves could assist the Royal Navy. In 1977 and 1978 the Squadron undertook periods of operational sea-training at Portland where the Navy prepared ships of the Fleet for the actualities of war. The drive to raise professional standards to what they had once been continued, and in 1979 the Senior Officer of the Squadron became for the first time a Royal Navy appointment. This proved an outstanding success providing much needed continuity and coordination and a permanent link with the ships of the Fleet in furtherance of the policy, successfully urged by the Mitchell Committee in 1974, that the Reserve is the part-time element of a single naval service. Also during 1979 London Division was allocated, once more, a coastal minesweeper, and by the end of that year most, if not all,

Contrasts in duty for HMS Glasserton. ABOVE *cheering HM Queen Elizabeth The Queen Mother on board HMY* Britannia *on completion of Royal Escort duty in 1980. (RN)* BELOW *ploughing through the Bay of Biscay on her way to the Mediterranean. (RN)*

Divisions had their own minesweepers again. By 1982 the Squadron was able to repeat the success it had achieved in 1970 by undertaking, once again, a N.A.T.O. operational exercise in the Mediterranean, working extensively with allied navies.

Recent years have seen increasing economic and social pressures on those who give up their time to undertake volunteer training. Employers are less ready than they were to recognise the advantages to their employees of voluntary service. At the

same time the integration of the Reserve with the Royal Navy, on a regular day-to-day basis, has inevitably led to an increase in paperwork and proliferation of naval authorities with which each Division has to deal. London, as the largest Division, found itself sorely pressed.

In order that management could be more evenly spread amongst members of the Division, and more readily carried out within the time-scale of a drill night, the Division was reorganised in February 1981 into three sub-divisions each with its own commanding officer in much the same way as the original ten companies with which the Division started in 1903. Every member, with the exception of Heads of Department, now belongs to *Hawke, Nelson* or *Hood* sub-divisions and new entries are allocated to *Anson*. These names commemorate the Battalions of the Royal Naval Division.

Only time will tell if this internal re-organisation, the most substantial since the Division re-started after the Second World War, meets the needs of the future. The fact that the Reserve continues to fulfil with spirit, determination and pride the many and various tasks it is asked to perform in support of the Royal Navy, is a remarkable facet of our national maritime heritage, and a fine tribute to the individual men and women who often, seemingly against all odds, continue, in naval parlance, to 'make it so'.

When Her Majesty the Queen travelled upstream by barge during the celebration of her Silver Jubilee in June 1977, she was cheered by the officers and men of London Division who manned the ship for the occasion. The customary three cheers were given, and done 'according to the book' ('Hip, Hip, Hip, Hurrah'). Within days, the Admiralty Board ordered that, for the forthcoming Review of the Fleet at Spithead and in future, three cheers in the Navy would be expressed by 'Hip, Hip, Hooray'. Could London Division have been instrumental in procuring a change of protocol? Apparently 'Hurrah' is literary and dignified, whereas 'Hooray' is the popular version, and only two 'Hips' are traditional. Thus, perhaps aided by London Division experience, the Navy has reverted to the traditional and popular version – a cheering thought indeed.

As the *President* is the only naval presence in the City of London, it has always had a significant part to play in keeping the Royal Navy in the public eye. In the early days before the First World War this took the form of organised 'March Outs' when the Division, with guard, band and a gun-carriage or two, would simply march through the streets on a pre-arranged route to remind the citizens of London they had a Navy. As already related, the annual Field-Gun Competition in Hyde Park attracted considerable public interest, many spectators and much favourable comment in the press. The Division's skill at field-gun drill was such that it provided public displays in Earl's

London Division marching in the Lord Mayor's Show passing the saluting base at Mansion House. (RN)

Court at the then equivalent of to-day's Royal Tournament. Between the wars, in the Twenties and Thirties, although the skills with the field-gun were lost, the annual Navy Weeks took their place as a means of focusing public attention on the Navy.

Since the Second World War, London has, together with all other Reserve establishments, maintained this tradition, and successive generations of Reservists throughout the country have, in addition to their training, played their part in furthering the cause of the Royal Navy in their local communities. In London in the Fifties and Sixties the 'Navy Week concept' of the Thirties was to some extent revived, and many will remember the Recruiting Weeks held each year in mid-summer when the drill ships were alive with naval activity, sheer-legs were rigged

between the stern of the *Chrysanthemum* and the bows of the *President*, and whalers and cutters were to be seen under way in King's Reach, watched by the crowds on the Embankment. In these years also, the Division took part in the annual Easter Parade in Battersea Park and provided the guard and colour party which traditionally began the Service Display at the annual Festival of Remembrance in the Albert Hall. Throughout the years there have been many moments when members of the Division were exposed to the gaze of a public which makes no distinction between 'part time' and 'full time' Navy. To them it was always simply a matter of 'the Navy's here'.

The earliest participation by the Division in the Lord Mayor's Show can be traced back to 1907. This annual event in London's calendar is an ever popular one, especially for children and young people. Recent highlights have included Sir Bernard Waley-Cohen's Lord Mayor's Show in 1960, when 'The Navy' was his theme, and Sir Robin Gillett's innovation of pausing and stepping from his coach for a welcome 'tot' in 1976 (see frontispiece). This traditional naval greeting has continued, and has now developed into a feature of the Lord Mayor's return

The Wardroom Mess HMS President *prepared for a mess dinner 1982.*

The gangway HMS President. (*Note hooks for hammocks – top left.*)

journey to the Mansion House, watched by thousands who throng the Embankment route.

It was at the first annual inspection of the Division in December 1904 that Admiral Rice expressed the hope that the City of London would show some practical token of its appreciation. This was perhaps too soon after the formation of the Division to expect the City to react. It is fair that time should have been allowed to run, so that the mettle of London's naval volunteers could be well and truly tested. The 'apprenticeship' of London's Navy which had lasted almost eighty years and encompassed two world wars, ended on 18 April 1983 when the Divisional Crest was mounted in Guildhall alongside those of the 'privileged' Regiments, in formal recognition of the Division as a City Volunteer Force.

The Divisional crest mounted in Guildhall, incorporating the ancient armorial ensigns of the City dating from 1381 when the red cross of St George was first combined upon a shield with the sword of St Paul, London's patron saint.

By the Captain

The Years Ahead

Although there have been periods in the history of the Naval Reserve when clearly the Royal Navy was uncertain how to use its enthusiastic volunteers, the same cannot be said today. By fully integrating the volunteers into the day-to-day running of the permanent service, the Reserve really is now the 'part-time element' of the Royal Navy. It is on immediate standby to fulfil a pre-determined role as soon as a crisis occurs, and is no longer a Reserve in the sense of 'standing behind' the regular forces.

This was demonstrated during the recent South Atlantic emergency when upwards of a hundred specialist Reservists voluntarily left their civilian jobs to undertake temporary duties: some to man maritime headquarters and communication centres; other to assist in the preparation of trawlers taken out of trade for mine counter-measures at Port Stanley. Also a number of medical officers provided support where needed. The fleet of merchant vessels, which sailed in support of the Task Force, could not have been mobilised so smoothly without the expert help of List 1 R.N.R. personnel who, with their intimate knowledge of the Merchant Service, volunteered to man posts ashore and even sail with the ships as part of the Force, the achievements of which are now enshrined in history.

There can be no doubt that the role played today by the Reserve is vital to national defence. London Division is well placed after the difficult years of the Seventies to move forward with confidence. The outlook is bright. The new generation of Fleet minesweepers is now being built, with the first ship due in service next year. The recent addition of the twenty-four knot Fairey Tracker for sea training is an exciting prospect which brings to mind the dashing image of those who served in Coastal Forces during World War II.

Our new headquarters, which will be about a mile downstream of King's Reach will give us the capacity to berth these new sea training craft freely alongside. As we therefore step ashore to start a new chapter in the continuing story of London Division, it is well for us to have in mind the traditions and accomplishments of past generations. With modern ships and new technology there will be much in the years ahead to absorb and fascinate the young volunteer interested in the sea and ships.

The need for a strong Navy has never been greater at any time in our history. London Division is proud to play its part in securing this end.

H.M.S. *President* March 1983 G.K.B.

Acknowledgements

London Division's gratitude to many people in the production of this book is immense.

His Royal Highness The Prince of Wales has conferred upon us the signal honour of contributing the Foreword.

Admiral of the Fleet Lord Lewin has very kindly written the Preface.

Gordon Taylor has given most generously of his time and skill in writing the text and unfolding so richly the story of our first eighty years.

It is impossible to list all who have contributed but we would like particularly to mention the late Lieut.-Commander Robert Bradley who started the project; Viscountess Boyd for access to material belonging to her father, the first Commanding Officer of London Division; former Commanding Officers, particularly Commodore Charles Noble, the late Commodore James Young and Captain Basil West; Captain Tony Sainsbury; Mr. Arthur Watts; Lieut.-Commander Leonard Tilsley; First Officer Mary Goodhart; Lieut.-Commander Gordon Connell; Lieut.-Commander Charles Dewey; Petty Officer Wren Pat Blackett; the editors of 'London Log', Lieut.-Commander John Baxter, Lieut.-Commander Maitland Thornton, who has also kindly provided material from his own collection, and Chief Petty Officer Jack O'Connor; Mrs. Louba Whitfield, Third Officer Sarah O'Connor, Mr. Bryan Mayson and Miss Fiona Spears for their help with the art work; and Mr. Colin Whyman and Mr. Julian Herbert for the photography.

We acknowledge with thanks the help of the Fishmongers' Company, the Naval Historical Library, the Naval Historical Branch, the National Maritime Museum, the Maritime Trust and the Guildhall Library.

We are indebted to those who have helped with the typing and to all who have contributed in so many ways.

This book is the culmination of ten years of collecting historical material and personal recollections from past members of the Division before they are lost to posterity. It is inevitable that things will have been left out and not everyone's recollections included, but perhaps those who have further material or memories of past years will be encouraged to let us have them for the archives which it is planned to establish in our new shore headquarters.

In publishing our story, we hope that, not only will our past be of interest, but the role which the Royal Naval Reserve plays today will be more clearly understood.

George Beattie
Peter Woodward
Lesly Trodd
Diana Mason

History Committee
HMS *President*
March 1983

Appendices

1

To the R.N.V.R.

A Goodly Heritage

Ronald A. Hopwood

In the days when the Press Gang went out on its beat
    To trap the unwary to serve in the Fleet
So fine was its mesh, and so craftily set
    That all sorts of fishes were trapped in the net,
Transmuted afloat by a magical touch
    Into smart upper yard men, and honoured as such.

By a just dispensation a Ship of the Line
    Bestowed, in return, a transfusion of brine
Which their offspring revealed in a love for the sea,
    So that butchers and bakers *et cetera* should be
Entitled, whatever their status or caste,
    To shiver their timbers with those of the past.

And so it fell out that the passage of years
    Gave birth to a body of sea volunteers,
From office and workshop, from garden and bench,
    Endowed with a fervour which nothing could quench,
Eschewing their evenings at theatre or bar,
    They started creating the R.N.V.R.

Unsung and unheeded, a breed of their own,
    Whom the Press (times have changed) left severely alone,
They recruited and drilled, and rejoiced at the sight
    When *President's* school overflowed every night,
Though their growth was unnoticed, their roots, all unseen,
    Were anchored deep down where their fathers' had been.

So the dreamers composing this Nautical Guild
    Laid the keel, well and true, on which others could build
When the war drums were rolling; and so as desired
    Was fashioned the craft which their vision inspired,
And the old upper yard men smiled down from afar
    As their scions flocked in to the R.N.V.R.

From City and Market, from Village and Town,
    From hillside and paddock they came to their own;
'Twixt Royals and Radar the gulf's pretty wide,
    But the R.N.V.R. took it well in their stride,
Leaving E-boats in doubt as they went to their fate
    If the stripes of their victors were wavy or straight.

Threadneedle Street experts, regardless of rank,
    Would steer for the Dogger instead of the Bank,
And farmers, declaring that tractors were vain,
    Were all for white horses when ploughing the main,
While Members of Parliament, fresh from the shore,
    Fell in at Divisions as never before!

And when it was over, for such are the rules,
    They all recommissioned the benches and stools,
But the salt in their blood set their hopes running high
    That their trades would keep better and sweeter thereby,
For they'd learn, if their wagons they'd hitch to a star,
    No tow rope's so strong as the R.N.V.R.

The Voyage of H.M.S. President

A Dream

Mr Punch means no disrespect to HMS President, *which being moored in the Thames off Bouverie Street, he had always looked upon as his guardship, but he has often wondered what would happen if only a few thousands of the officers and men borne on her books were to issue from the Admiralty and elsewhere – but especially from the Admiralty – and go on board their ship; hence the disquieting dream that follows.*

It was eighteen bells in the larboard watch with a neap-tide running free,
And a gale blew out of the Ludgate Hills when the *President* put to sea;
An old mule came down Bouverie Street to give her a helping hand,
And I didn't think much of the ship as such, but the crew was something grand.

The bo'sun stood on a Hoxton bus and blew the Luncheon Call,
And the ship's crew came from the four wide winds, but chiefly from Whitehall;
They came like the sand on a wind-swept strand, like shots from a Maxim gun,
And the old mule stood with the tow-rope on and said, "It can't be done."

With a glitter of wiggly braid they came, with a clatter of forms and files
The little A.P's they swarmed like bees, the Commodores stretched from miles;
Post-Captains came with hats in flame, and Admirals by the ell,
And which of the lot was the biggest pot there was never a man could tell.

They choked the staggering quarter-deck and did the thing no good;
They hung like tars on the mizzen-spars (or those of the crowd that could);
Far out of view still streamed the queue when the moke said, "Well I'm blowed
If I'll compete with the 'ole damn Fleet," and he pushed off down the road.

And the great ship she sailed after him, though the Lord knows how she did,
With her gunwales getting a terrible wetting and a brace of her stern sheets hid,
When up and spoke a sailor-bloke and he said, "It strikes me queer,
And I've sailed the sea in the R.N.V. this five-and-forty year.

"But a ship as can't 'old 'arf 'er crew, why, what sort of a ship is 'er?
And oo's in charge of the poor old barge if dangers do occur?
And I says to you, I says ''Eave to, until this point's agreed',"
And some said "Why?" and the rest "Ay, Ay," but the mule he paid no heed.

So the old beast hauled and the Admirals bawled and the crew they fought like cats,
And the ship went dropping along past Wapping and down by the Plumstead Flats;
But the rest of the horde that wasn't aboard they trotted along the bank,
Or jumped like frogs from the Isle of Dogs, or fell in the stream and sank.

But while they went by the coast of Kent up spoke an aged tar –
"A joke's a Joke, but this 'ere moke is going a bit too far;
I can tell by the motion we're nearing the ocean – and that's too far for me;"
But just as he spoke the tow rope broke and the ship sailed out to sea.

And somewhere out on the deep, no doubt, they probe the problems through
Of who's in charge of the poor old barge and what they ought to do;

And the great files flash and the dockets crash and the ink-wells smoke like
 sin,
But many a U-boat tells the tale how the *President* did her in.

For many have tried to pierce her hide and flung torpedoes at her,
But the vessel, they found, was barraged round with a mile of paper matter;
The whole sea swarms with Office Forms and the U-boats stick like glue,
So nothing can touch the *President* much, for nothing at all gets through.

But never, alack, will she come back, for the *President* she's stuck too.

A.P.H.
Punch May 15th 1918

Note: From May 1878 officers serving in the Admiralty and at the Royal Naval College Greenwich were borne on the books of the President. This has always caused confusion with 'President R.N.R.'. This may shortly be resolved by changing the name of the Royal Navy's London establishment. The name President will be retained by the Royal Naval Reserve.

Headquarters

God gave all men all earth to know,
 But since men's minds are small,
Ordained to each one spot should
 grow
 Familiar over all.
So, though I never made the choice,
 My lot has fallen to me
In a dreary place—Headquarters,
 Commercial Road, S.E.

Let some their Baltic pines content,
 And one his Sussex Down,
Heaven gives small grace of
 preferment
 To him who lives in Town.
Choose ye your need from Thames
 to Tweed,
 No choice was given to me,
The Fates ordain that I remain
 In Prince's Wharf, S.E.

No garden fair, no wooded kloof,
 No foliage spreads o'er
Its corrugated iron roof,
 Its grime-swept concrete floor.
Bare space, like some dead castle
 hall,
 Where, phthisis following fast,
Our cracked and crazy sheet-iron
 wall
 Lets in each freezing blast.

Clean of all hint of elegance,
 Clear-flung, without a check,
The vision leaps the dim distance
 From door to battery-deck.
Six-inch, four-inch, and four-point-
 seven—
 Black Down and Beachy Head
May lift their furrowed brows to
 Heaven,
 Fate gives me these instead.

Here nightly throng the eager hearts,
 Enthusiastic still,
Each with the murk of city marts
 To drive away in drill.
Here Muckle tells us that he likes
 —In case we do not know—
To have his borrowed marling-spikes
 Returned before we go.

Here beetle-browed Bill Avery
 stands,
 To all the troops fore-square
And loudly yaps his curt commands,
 His 't'shun!' and 'as you were!'
Or, haply, in the office he
 To some poor shivering youth
(His rifle not returned, may be)
 Preaches Heaven's wholesome
 truth.

Here, in this office, Perkins wields
 —Far mightier than the sword—
The full-charged pen; and each man
 yields
 Who reads his written word.
Here through the strong
 unhampered day,
 The tinkling silence thrills;
In peace—while Perky's cross the
 way—
 George Colbert books up drills.

We have no pleasance to delight,
  Small meed have we of cheer,
Only the canteen on your right
  And Worthington's best beer.
Bare barn-like room wherein we
    feast,
  Where, baleful, Leeson glares
While Jock and Marlow and Bill
    East
  Break up the canteen chairs.

See now our sea-forgotten walls,
  Our wharf of stranded pride,
Mark where some reeking sewer
    crawls
  To find the fickle tide.
Here is our frontage to the stream,
  Here hoist we in our boats,
Hauling on luffs hooked to the beam,
  To Barrett's strident notes.

Here leaps ashore the full Thames
    tide
  And, borne upon its brink,
Our blighted battered barges ride
  And haply swim—or sink.
And here the Thames fogs lap and
    cling,
  Here deep at foot there lies
Thames mud that smells like
    anything
  But dawn in Paradise.

I will look North towards the Strand
  Where, towering spire on spire,
Temple and Fane and Fabric stand
  For Hatred, Fear, Desire.
Or East, where blunt-nosed lighters
    slow
  Glide upward from the Pool,
West, where stands, bathed in sunset
    glow,
  Man's Palace of Mis-rule.

Or South—one blank and sightless
    wall
  Stares blindly down on me,
The Naval Volunteer Drill Hall,
  Commercial Road, S.E. . . .
Take thou thy meed of service given,
  Tithe of my years thy toll!
Thy corrugated iron has driven
  Deep down into my soul!

God gave all men all earth to love,
  Since His Elect are small
To them perchance one spot may
    prove
  Beloved over all.
So I rejoice—I do *not* think!—
  That all my nights should be
Spent in this frigid concrete sink,
  Headquarters, R.N.V.

From *My 'Ditty' Box*
C. G. Murray C.P.O. R.N.V.R.
1913

Divisional Song of the Royal Naval Volunteers

(London Version)

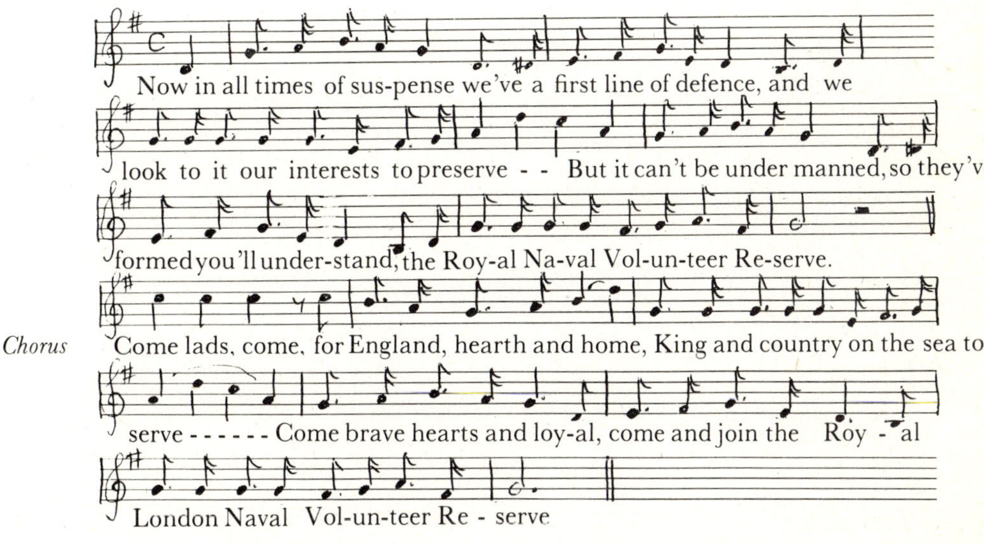

Now in all times of suspense, we've a first line of defence,
And we look to it our interests to preserve,
But it can't be undermanned, so they've formed, you'll understand,
The Royal Naval Volunteer Reserve.

From the little Maxim gun, to the heavier 6-inch one,
We are learning all the details through and through;
So wherever we may be the world will always see,
That the Volunteers are firing straight and true.

So now the Force has grown, we'll let it well be known,
That the Volunteers won't from their purpose swerve;
So efficient every man, for no Corps is finer than
The Royal Naval Volunteer Reserve.

But if any Foreign Fleet with contempt our Flag should treat,
And the music of our warships has to play,
Then the Naval Volunteer won't be standing in the rear
But will lend a hand to help to win the day.

So when all the fighting's done, and the battle's lost and won
And when volleys three are fired towards the sky,
Then men shall witness bear, that the Naval Volunteer
Knows not only how to fight, but how to die.

Well equipped against the foe, all the Volunteers will go,
And they'll wipe him off the face of all the Seas;
Then he'll very soon find out what's the fate of those who flout
The Flag that braves the battle and the breeze.

— *Durnford.*

The Ballad of Codson's Beard

I'LL tell you a yarn of a sailor-man with a face more fierce than fair
Who got round that on the Navy's plan by hiding it all with hair;
He was one of a hard old sailor-breed and had lived his life at sea,
But he took to the beach at the nation's need and fought with the R.N.D.

Now Brigadier-General Blank's Brigade was tidy and neat and trim,
And the sight of a beard on *his* parade was a bit too much for him.
"What is that," said he with a terrible oath, "of all that is wild and weird?"
And the Staff replied, "A curious growth, but it looks very like a beard."

And the General said, "I have seen six wars and many a ghastly sight,
Fellows with locks that gave one shocks and buttons none too bright,
But never a man in *my* Brigade with a face all fringed with fur;
And you'll toddle away and shave to-day"—but Codson said, "*You err.*

"For I don't go much on wars as such, and living with rats and worms,
And you ought to be glad of a sailor-lad on any old kind of terms;
While this old beard of which you're skeered it stands for a lot to me,
For the great North gales and the sharks and whales and the smell of the dear grey sea."

New Generals crowded to the spot and urged him to behave,
But Codson said, "You talk a lot, but can you *make* me shave?
For the Navy allows a beard at the bows, and a beard is the sign for me
That the world may know, wherever I go, I belong to the King's Navee."

They gave him posts in distant parts, where few might see his face,
Town Major jobs that break men's hearts and billets at the Base;
But whenever he knew a fight was due he hurried there by train,
And when he'd done for every Hun they sent him off again.

Then up and spake an old sailor, "It seems you can't 'ave 'eared,
Begging your pardon, General Blank, the *reason* of this same beard;
It's a kind of a sart of a *camyflarge*, and that I take to mean
A thing as 'ides some other thing wot oughtn't to be seen.

"And I've brought you this 'ere photergraph of wot 'e *used* to be
Afore he stuck that fluffy muck about 'is phyzogmy."
The General looked and, fainting, cried, "The situation's grave,
The beard was bad, but, *Kamerad!* he simply must not shave!"

And now, when the thin lines bulge and sag and man goes down to man,
A great black beard like a pirate's flag flies ever in the van;
And I've fought in many a red-hot spot where death was the least men feared,
But I never saw anything quite so hot as the Battle of Codson's Beard.

A. P. H.

The figurehead of the 5th President *now in Fishmongers' Hall and restored by Clive Lee of the City and Guilds Art School, Kennington.*

Ships bearing the name of HMS PRESIDENT

Type	Name held from ___ to ___		Remarks
1st 20 gun	Built	1646	The 'Little' or 'Old'
2nd 42 gun	Built	1649	The 'Great'
		1660	Renamed *Bonaventure*
3rd 5th rate	prize	1806	ex *Piedmontaise*
		1815	broken up
4th 5th rate	prize	1815	
		1818	broken up
5th 4th rate*	Built	1829	Reserve drill ship 1861
		1903	Sold
6th sloop		1903	ex *Gannet* (built 1878) renamed
		1911	*President* lent to C. B. Fry, renamed *Mercury*. (Renamed *Gannet* when acquired by the Maritime Trust in 1971)
7th sloop		1911	ex *Buzzard* R.N.V.R. drill ship
8th sloop		1921	ex *Saxifrage* R.N.V.R. drill ship

* *Figurehead came from 5th.*

Battle honours of ships bearing the name of HMS PRESIDENT

Prepared by Lieut.-Commander W. M. Thornton

Portland Feb. 18–20 1653 (First Dutch War) Second ship of the name built at Deptford in 1649. 4th rate, 445 tons, 42 guns, 130 officers and men (renamed Bonaventure in 1660). Captain was Captain Graves R.N. Dutch Admiral Tromp with 80 warships was escorting 200 merchantmen bound for Holland. As he passed Portland he was attacked by 80 British ships under Blake. The convoy scattered towards France and the action finished off Cape Gris Nez. The Dutch lost 8 warships and 30–50 merchantmen were captured. The British fleet had one ship sunk and 3 disabled. *President* supported main fleet and acted as a scout.

Gabbard June 2–16 1653. Same ship as before. Action commenced off Gabart Sand near Thames Estuary. Dutch fleet under Tromp prepared to attack British shipping off English coast. He had 98 ships and 6 fireships. The British fleet under Blake and Sea Generals Monck and Deane (who was killed) and Admirals Penn and Lawson attacked the Dutch with 100 ships and 5 fireships. The Dutch were swept back to their own coast, and as the action lasted a number of days the opposing fleets often lost contact. *President* with other similar ships scouting ahead of the main fleet sighted the main enemy after their withdrawal between Ostend and Blankenburg. The Dutch lost 11 ships captured and 8 destroyed. No British ships were lost but casualties were heavy. The action resulted in the closing of the Dutch Ports for the remainder of the war.

Scheveningen July 28, 1653. Same ship as above. In this action as before *President* scouted ahead of the main fleet seeking intelligence data. British fleet under Sea Generals Monck and Jordan attacked 38 Dutch ships sailing towards Scheveningen. Contact was lost because of bad weather but on July 30 the main Dutch fleet of 120 ships under Tromp engaged the British fleet of 90 warships and 20 auxiliaries 1½ miles off Scheveningen. The real action did not start until the next day and resulted in the bloodiest engagement of the war, lasting from 7 a.m. to 8 p.m. Tromp was killed by a musket ball and Jan Evertsen took command of the Dutch. The Dutch lost 14 ships though the British claimed 20–30. The British losses consisted of 2 ships and 1 fireship. Casualties were high. The result of the action was the final crushing of Dutch sea-power for the remaining year of the war.

Java May 23–Sept. 18, 1811. *President* was a 5th rate (frigate) 3rd ship of the name (ex-French *Presidente* built 1801 and captured in the Bay of Biscay on Sept. 27, 1806 by H.M.S. *Dispatch*). She was of 1,152 tons, 38 guns and carried 284 officers and men. (She was later renamed *Piemontaise* and was scrapped in 1815.) Her captain during the reduction of Java was Captain Samuel Warren R.N. British fleet of 4 sail-of-line, 14 frigates, 7 sloops plus 8 H.E.I. Co.'s cruisers under Rear Admiral The Hon. R. Stopford (up to Aug. 9 Commodore Broughton) along with army personnel attacked Franco-Dutch forces in Java. Action consisted mainly of land fighting but included a naval brigade at storming of Cornelius. Ships were used for bombardments and a number of small sea actions were fought. On Sept 12 *President* in company with *Nisus* and *Phoebe* received the surrender of the seaport of Cheribon. Result of action, Java became British on Sept. 18, 1811.

San Sebastian Aug. 10–Sept. 8, 1813. Same ship as above but under the command of Captain George Tobin R.N. British squadron of 12 ships, 2 schooners, 1 cutter and 2 gunboats under command of Sir George R. Collier, co-operated with General Graham's Anglo-Spanish force against the French in the reduction of San Sebastian. Seamen were landed from fleet and 2 divisions of ships boats were used in occupying the town. The action ended with the capture of the citadel on Sept. 8.

1907

No. 1 Company

Lieutenant – Edmund Wildy.
Sub-Lieutenant – Frederick James Paice.
Midshipman – Claude Bernard Rowe.
Surgeon – Francis John Hannan, M.D.
C.P.O. Instructor – C. Knowlson.
Chief Petty Officer – A. G. Josling.

Petty Officers –	4
Leading Seamen –	4
Men –	65
Total Strength –	***79***

Date of Formation of Company,
November 10th, 1903.
Drill every Tuesday, 6.30 to 8.30 p.m.

No. 2 Company

Lieutenant – Hugh Burdett Money-Coutts.
Sub-Lieutenant – Alan Hilary Moore.
Midshipman – E. P. Evans.
Surgeon – Marcus S. Paterson, M.B.
C.P.O. Instructor – J. Walters.
Chief Petty Officer – F. A. Spiers.

Petty Officers –	4
Leading Seamen –	4
Men –	52
Total Strength –	***66***

Date of Formation of Company,
November 26th, 1903.
Drill every Monday, 6.30 to 8.30 p.m.

No. 3 Company

Lieutenant – Thomas H. Roberts-Wray.
Sub-Lieutenant – Frank Risch.
Midshipman – John Stuart Knill.
Surgeon – Frank Wybourn Smith.
C.P.O. Instructor – H. Pickard.
Chief Petty Officer – O. C. Beale.

Petty Officers –	4
Leading Seamen –	4
Men –	65
Total Strength –	***79***

Date of Formation of Company,
December 10th, 1903.
Drill every Thursday, 6.30 to 8.30 p.m.

No. 4 Company

Lieutenant – William James Fernie.
Sub-Lieutenant – Sydney Searle.
Midshipman – Eric Worsley Gandy.
Surgeon – Arthur R. Brailey, M.B.
C.P.O. Instructor – W. Fenwick.
Chief Petty Officer – J. Crowley.

Petty Officers –	4
Leading Seamen –	4
Men –	61
Total Strength –	***75***

Date of Formation of Company,
December 17th, 1903.
Drill every Monday, 6.30 to 8.30 p.m.

No. 5 Company

Lieutenant – Arthur Cecil Curtis.
Sub-Lieutenant – William Richard Macdonald.
Midshipman – Roger Charles Anderson.
Surgeon – Reginald J. E. Hanson, M.B.
C.P.O. Instructor – A. Gomer.
Chief Petty Officer – T. Lelliot.

Petty Officers –	4
Leading Seamen –	4
Men –	40
Total Strength –	***54***

Date of Formation of Company,
December 20th, 1903.
Drill every Friday, 6.30 to 8.30 p.m.

No. 6 Company

Lieutenant – Charles Devereux Marshall.
Sub-Lieutenant – William James Hampsheir.
Midshipman – Frank William Hollingsworth.
Surgeon – Edward J. Steegman, M.B.
C.P.O. Instructor – W. H. Shute.
Chief Petty Officer – W. Glegg-Smith.

Petty Officers –	4
Leading Seamen –	4
Men –	44
Total Strength –	***58***

Date of Formation of Company,
January 5th, 1904.
Drill every Tuesday, 6.30 to 8.30 p.m.

No. 7 Company

Lieutenant – Oswald Hesketh Hanson.

Sub-Lieutenants –
John R. Kemeys Warneford.
Ralph Martin Soames.
Midshipman –
Surgeon – Herbert Robert Power.
C.P.O. Instructor – D. Harris.
Chief Petty Officer – G. A. W. Trinder.

Petty Officers –	4
Leading Seamen –	4
Men –	52
Total Strength –	***66***

Date of Formation of Company, January 7th, 1904.
Drill every Thursday, 6.30 to 8.30 p.m.

No. 8 Company

Lieutenant – Alexander Lyon.

Sub-Lieutenants –
The Hon. S. G. J. Macdonald.
Algernon C. H. Drummond.
Midshipman –
Surgeon – James John Marsh, M.D.
C.P.O. Instructor – S. Munro.
Chief Petty Officer – A. J. Evans.

Petty Officers –	4
Leading Seamen –	4
Men –	57
Total Strength –	***71***

Date of Formation of Company, January 15th, 1904.
Drill every Friday, 6.30 to 8.30 p.m.

No. 9 Company

Lieutenant – Henry Douglas King.
Sub-Lieutenant – Bernard T. Monier-Williams.
Midshipman –
Surgeon – Arthur Douglas Cowburn.
C.P.O. Instructor – A. Newbery.
Chief Petty Officer – A. H. Ninnis.

Petty Officers –	4
Leading Seamen –	4
Men –	60
Total Strength –	***73***

Date of Formation of Company, May 6th, 1904.
Drill every Wednesday, 6.30 to 8.30 p.m.

No. 10 Company

Lieutenant – Viscount Curzon.
Sub-Lieutenant – William Ginman.
Midshipman – Hubert Alington Yockney.
Surgeon – James Keogh Murphy, FRCS, MD, MA.
C.P.O. Instructor – F. Pulford.
Chief Petty Officer – R. Say.

Petty Officers –	4
Leading Seamen –	4
Men –	78
Total Strength –	***92***

Date of Formation of Company, September 7th, 1904.
Drill every Wednesday, 6.30 to 8.30 p.m.

The above is taken from the booklet prepared for the visit of the Prince of Wales in 1907.

2

Captain The Earl of Iveagh KG CB CMG VD FRS RNVR

(formerly The Hon. Rupert Guinness)

By the time he was appointed to be London Division's first Commanding Officer at the age of 29 in 1903, Rupert Guinness was already a well known figure, mainly through his prowess as a sportsman and through his increasing involvement in public affairs.

In 1892 he won the school sculling championship and in the following year he rowed in the Eton eight which won the Ladies Plate at Henley. On leaving school he set out to excel at rowing and did so by winning in 1895 both the Diamond Sculls and the Wingfield Sculls, thus becoming the amateur Sculling champion of Great Britain. This brought him instant fame, but the 'blue' which assuredly would have been his when he went up to Trinity College Cambridge was denied him as his father forbade him to continue rowing believing, mistakenly, that he had a strained heart. So Guinness took up yachting and again achieved success by winning the King's Cup in his debut in 1901. In his yacht 'Leander' he beat the Emperor of Germany's 'Meteor' and the King's former yacht 'Britannia', skippered at the time by Sir Richard Williams-Bulkeley, who was to become the first Commanding Officer of Mersey Division.

Rupert Guinness, the eldest son of the first Lord Iveagh, was born in London in 1874 in the same year as Winston Churchill with whom he was to cross paths – and swords – sharing a governess in Dublin as five-year-olds. The Guinness family lived in Dublin where young Winston accompanied his father who was ADC to his grandfather the Duke of Marlborough, then Viceroy of Ireland. At the end of their lives Guinness and Churchill walked together in the Garter Procession at Windsor. Guinness was made a Knight of the Garter, a personal gift of the Sovereign, in 1955 at the age of 81. This was not his last distinction for he was to receive at the age of 90 the rare honour of being elected a Fellow of the Royal Society for his contribution to agriculture. In his modest way he declared he could not understand why these honours had been given to him, but few things, nevertheless, delighted him more.

1903 was an eventful year for Rupert Guinness. Not only did he take over as Commanding Officer of the newly formed London Division, he was also adopted as Unionist candidate for the East End constituency of Haggerston and, above all, when in the October he married Lady Gwendolen Onslow he found the ideal partner in a long and happy marriage. At a by-election in 1908 Guinness won the Haggerston seat after a hard election campaign (the voting figures are interesting compared with today – Tories 2867, Liberals 1724 and Socialists 986). Lady Gwendolen organised campaign helpers, among whom were members of the R.N.V.R. who were given the job of tracing those on the electoral roll who had moved house. It was reported that on the night of his victory shoals of empty Guinness bottles were found floating near Waterloo Bridge thought to have been thrown overboard from the *Buzzard*. Guinness lost the seat in the General Election of 1910 but was later returned as member for Southend, the constituency he represented until he took his seat in the House of Lords on his father's death in 1927.

Rupert Guinness proved a dedicated and resourceful leader. He gave generously of his time and money, spending most evenings, when not in the House of Commons, on Divisional matters. The Globe's verdict in 1920 that 'The R.N.V.R. (in London) owed almost everything to the splendid work he did for it in pre-war days' was scarcely an exaggeration.

Captain Edmund Wildy OBE VD RNVR

Volunteering came quite naturally to Edmund Wildy (pronounced Will dy).

At the age of 16 he joined the famous Rifle Volunteers (The Artists) as a bugler and soon attained the position of Bugler Major organising a bugle band which was the pride of the regiment. As a keen yachtsman, however, he was attracted to the Royal Naval Artillery Volunteers and he enrolled in that corps as a seaman (then called a gunner) and after a few years commanded No 5 Battery.

When the R.N.A.V. was disbanded Wildy was very disheartened (rumour had it he grew a moustache), but he remained determined if possible to revive the old movement, keeping in touch with a small group of men who felt the same and actively canvassing for the re-formation of a naval volunteer force. His efforts were rewarded when he was asked to join the committee which eventually recommended the setting up of the Royal Naval Volunteer Reserve. This objective achieved, Wildy then turned his attention to finding recruits. He issued a memorandum to members of the Stock Exchange giving 'the preliminary conditions for the formation of a naval volunteer corps' and stating at the end that 'further information can be obtained from and application to join be made to – E. Wildy, Swan Chambers, Copthall Avenue, E.C.'

In 1903 Wildy was offered the first commission as a Company Officer in London Division and appointed in command of No 1 Company, thereby becoming second-in-command of the Division under Rupert Guinness. Much of the day to day running of the Division would have fallen to him and he must have found the experience he gained in the R.N.A.V. invaluable. The numerous orders and minutes Wildy issued showed him to be professional and to have had an eye for detail. His influence would have affected all Divisional activities – training, ceremonial and social. No doubt, among other things, Wildy was responsible for the formation of the bugle and drum band. At the annual dinner of No 1 Company in 1906 he sang 'Bold Dan'l' which must have been a favourite party piece of his for he is recorded as having performed it fifteen years earlier, at a 'sing song' in the *Frolic* while a CPO in the R.N.A.V.

During the war Wildy was appointed Drafting Officer at Crystal Palace and became a Commander. On the re-formation of London Division in 1921 he was Executive Officer for a year, thus ensuring the Division was again set up on the right lines. He retired with the rank of Captain.

A yachtsman of some repute, Wildy was a member of several well known yacht clubs, including The Royal Corinthian, The Royal Thames and The Royal Cruising. He never married and is remembered in the Twenties by his great nephew, The Right Reverend David Say, now Bishop of Rochester, then a small boy, as 'rather a formidable old gentleman'.

The R.N.V.R. owes its existence to the devotion and persistence of people like Edmund Wildy and London Division owes to him a debt of gratitude for his dedication and guidance in the early years.

ROYAL NAVAL VOLUNTEER RESERVE

EASTER ORDERS

The advance party will muster at Headquarters at 8.30 a.m., on Thursday, April 12th, and having stowed Ammunition, adaptors and tubes in limbers, will march at 9 a.m. for London Bridge Station.

Uniform.—Working rig, full equipment, Rifles and Bayonets.

Men will make their own arrangements for lunch.

The main body will muster at Victoria Station at 7.10 p.m., and will fall in by Companies. Train starts at 7.50 p.m.

Uniform.—Working rig, full equipment, Rifles and Bayonets.

Kit Bags to be brought to the Station, or sent off addressed to the Officer Commanding R. N. Volunteers, Cloak Room, Eastbourne Station, not later than Tuesday night, April 10th, so as to be there on the arrival of the main body. A cart will meet the train to take the Kits up to Quarters at the New College, Eastbourne.

All Officers and Men should have a good meal before going down, or provide sandwiches to eat in the train, as it will not be possible to arrange for a meal on arrival.

Men unable to go down with main body must follow on by later trains, and report on arrival.

Kit Bags.—No. and Name to be painted on outside:

123
A. SMITH,
No. 1 Co.,
R. N. V.

in letters 1½ inches long.

Men should carry in Kit Bags; No. 1 rig — White Working rig — spare pair of Boots — Slippers or Gymnastic Shoes — change of Underclothing and Socks — Sleeping Kit (Flannels and Sweater recommended) — Towel, Soap, etc. — Toilet requisites — Cleaning Kit — Boot Brushes, Blacking, and Brown Polish for equipment — Reefer Coat optional.

A mattress and two blankets will be provided for each man.

Officers to muster in No. 5 rig, with Overcoats (unless too warm), Swords and Gaiters, Brown Gloves.

Kit to be taken by Officers.

Nos. 4, 5 and 7, White Gloves and Field Leathers.

Officers are advised to provide themselves with Towel, Soap, etc., and spare pairs of Boots.

Officers will be billeted at the HYDRO HOTEL.

By order,
E. WILDY,
Lieutenant commanding No. 1 Company, R.N.V.

Richard Say, Geoffrey McDonell, John Cordingley
(The C.P.O.'s of the early years)

To have been the Chief Petty Officer of a Company before the First War would have required considerable dedication, for the job would have been both responsible and demanding. With the high degree of competition between the Companies, the training at the drill hall and in the *Buzzard*, the numerous parades and the weekend activities, there would have been much to do. Judging from the stories of the conditions of both Headquarters and the *Buzzard*, it could not have been a comfortable life and there would have been no prospect of promotion (it took a war to change this). Yet men of the calibre of John Cordingley, John Hemsted, Geoffrey McDonell and Richard Say were not only prepared to do it, but from all accounts loved it.

Commander Richard Say

Richard Say, who had been the Chief Petty Officer of No. 10 Company under Viscount Curzon, married a niece of Edmund Wildy having first met her at an R.N.V.R. dance. Say served at the Admiralty during the First War and ended up as a Lieutenant Commander with an OBE. Although well into his 60's at the outbreak of the Second World War, his services were accepted at the office of the Second Sea Lord where he gained a third stripe, finally taking off his uniform in 1946 at the age of seventy.

But for the fact he was already a member of the King's Colonial Yeomanry, a territorial regiment, Geoffrey McDonell would have been the first rating recruit of London Division, an honour which fell to Petty Officer Stanley Stocker. By the time McDonell had bought himself out of the Territorial Army he became No. 787 although he attended drills from the start. Knowing him to have been in the R.N.A.V., Edmund Wildy came to him in the Stock Exchange and asked him to join the new R.N.V.R., telling him that the Admiralty had stipulated that provided 200 recruits could be found London Division could be formed. The memorandum that Wildy circulated and his personal approach soon brought forth the necessary number from the Stock Exchange alone and they became No. 1 Company. McDonell became the Chief Petty Officer. Much of the organisation of the Easter Training weekends was done by him and in recognition of this he was given a massive silver loving cup bearing the inscription 'R.N.V.R. – C.P.O. McDonell (No. 1 Company) Eastbourne, Easter 1905. Presented by the Battalion'. 'The Battalion' consisted of men from every Company and from Sussex Division who had taken part. The Naval and Military Record reports that 'the presentation was made by C.P.O. Crowley (No. 4 Company).' After three years McDonell left to go to Canada. There, at the outbreak of the First War, he joined the Princess Patricia's Canadian Light Infantry as a private, later coming over to fight in Europe. By the time the war finished he had reached the rank of Lieutenant Colonel.

John Cordingley later followed McDonell as the C.P.O. of No. 1 Company and became renowned for, among other things, his Field Gun Displays. In addition to those given before the First Lord of the Admiralty, Mr Winston Churchill in 1912 and the First Sea Lord, Prince Louis of Battenberg in 1913, there are accounts of other displays which received commendation. On the 1st April 1912 Cordingley received a letter from the London Rifle Brigade thanking him and his crew for taking part in their 40th Assault at Arms. An extract from the letter reads '. . . I know I told you that we looked upon the R.N.V.R. to provide the Star Turn of the evening, but the Display was far and away above everything I had hoped for, and it is quite impossible for me to attempt to describe how magnificent we all thought your Field Gun Drill was . . .' The Daily Telegraph of the 19th August 1913 reported on the 'London Naval Volunteers' appearance at the forerunner of today's Royal Tournament, as follows:-

Air Vice-Marshal Sir John Cordingley

'An attraction of an exceedingly interesting character had been added to this week's programme at the Imperial Service Exhibition at Earl's Court. It consists of a display by about 200 members of the Royal Naval Volunteer Reserve . . . In the naval 12-pounder Field Gun Drill a battery in the charge of Chief Petty Officers Wright and Cordingley particularly distinguished themselves . . .'

Chief Petty Officer J. C. Wright had other claims to fame. Together with Chief Petty Officer C. G. Murray, whose verses were published in 'My Ditty Box' (see appendix) he produced a 'Grand Topical Revue' by Messrs. 'Merry and Bright'. They had as their musical director Seaman Foort, who was to become well known as Reginald Foort, the theatre organist. So there was no shortage of talent and most aspects of Divisional life were set to music including the popular Field Gun Display by No. 1 Company. Thus was born the first of a series of revues and concert parties given over the years. Between the wars a group called 'The Pirates' put on several performances and, since the Second World War, Pantomine has formed a continuing feature of Divisional life when time allows.

Captain Thomas H. Roberts-Wray, CB OBE VD RNVR

Roberts-Wray was in the R.N.V.R. from the beginning. He had actively campaigned for the resuscitation of a Naval Volunteer Service and became in 1903 the Lieutenant in command of No. 3 Company. A letter referring to annual training in the *Britannia,* in the archives of those days reads as follows:-

'Dear Roberts-Wray,

Now that the first lot of your men have left I hasten to give my account of them. I can truthfully say that they are the best lot we have had and that they are noticeably smart and intelligent. Everybody has noticed it, and incidentally they have been of great assistance to me during provisioning ship etc. when one watch was on leave. I think you may congratulate yourself on your "braves".

Yours sincerely,
William R. W. Kettlewell
Lieutenant R.N.'

Commander Thomas Roberts-Wray

In the period up to the outbreak of the First War Roberts-Wray, judging from his letters, was very much in command not only of his own company but of other events within the Division. As a successful engineer in civilian life he was concerned with the maintenance of the drill ship and drill hall, and his letters portray him as a man of considerable ability and confidence. In 1911 on seeing a procession of strikers marching through the streets of the City of London on their way to Tower Hill and noticing amongst their ranks a 'stoker blue jacket of the British Navy' he felt compelled to write to the Admiral Commanding Reserves drawing his attention to this matter. He was indefatigable in his efforts to improve the efficiency and popularity of the R.N.V.R. and was a keen yachtsman.

During the war Roberts-Wray was the senior staff officer at Crystal Palace and had in his charge at one time some 10,000 men who, according to Commodore The Duke of Montrose (ex Captain of Clyde Division) 'never gave him any trouble'. The Duke had known Roberts-Wray since 1903 and wrote a personal tribute to him on his death in 1943. He said of him '. . . Whenever there was R.N.V.R. work to do he was there with his happy smile and his strong voice. He had a fine personality. He was a splendid friend to all, and the very fact that so many young men in the past, and today, have the joy and privilege of serving their country with the Navy is a debt of gratitude they owe to Roberts-Wray.'

Lieutenant H. D. King

Commodore The Rt. Hon. Henry Douglas King CB CBE DSO VD MP RNVR

It is unfortunate that Commodore King is probably best remembered in London Division for the fact that he lost his life when the yacht 'Islander' foundered off the Cornish coast in 1930, for he had been a prominent member for over twenty-five years and had been in command for nine. But his sudden death was a considerable shock to the Division and at his funeral in London ratings were pall bearers while officers lined the path to the Church, bowing their heads in respect. The ship's company presented to London Division a solid silver bell in his memory which, until recently, when prudence dictated it should be securely locked away, hung on the gangway of the *President* near the entrance.

King was in the merchant navy until 1899 becoming a midshipman R.N.R. in 1893. He left the sea to take up farming, studied law and was called to the Bar (Middle Temple) in 1905, the year after he was commissioned in the R.N.V.R.

In August 1914 on the formation of the Royal Naval Division he was appointed to Drake Battalion which he commanded for a year in Gallipolli and France. King returned unscathed from Antwerp and arrived in the Dardanelles on the day Tisdall won his VC, 25th April 1915. He remained in Gallipolli until the evacuation in January 1916 by which time he had won a DSO, been thrice mentioned in despatches and been wounded. For the remainder of the war he served afloat and with Royal Naval siege guns in Flanders where he was awarded the Croix de Guerre.

Returning from the war, King was chosen to succeed Rupert Guinness and has the distinction of being the only commanding officer of London Division ever to be a Commodore-in-Command. He succeeded in the difficult task of re-establishing the Division and rekindling the volunteer spirit in the apathetic post-war climate of the 1920's.

Like Guinness, King was a Member of Parliament, representing North Norfolk as an Independent from 1918 to 1922 and South Paddington as a Unionist from 1922 until his death.

King was Parliamentary Private Secretary to Sir Hamer Greenwood, Chief Secretary for Ireland, before becoming a Conservative Whip and later held office as Lord Commissioner of the Treasury, Financial Secretary to the War Office and lastly as Secretary for Mines until a year before his death.

Captain Norman ffolliott Wells OBE VD RNR

Captain N. ff. Wells

In 1921 Wells succeeded Wildy as executive officer, holding the appointment for nine years, until he suddenly had to assume command in 1930 after Commodore King was drowned in a yachting accident. Three years later he himself had to relinquish his command unexpectedly when a severe heart attack prevented his carrying on.

Norman Wells came from a naval family – his father and brother were both admirals – and his great-grandfather, Admiral Thomas Wells, was one of the canopy bearers at Nelson's funeral.

Wells joined London Division in 1910 as a midshipman, being promoted to Lieutenant in 1913 in No. 6 Company. On the outbreak of war he was sent to the Royal Naval Division and went to Belgium. He escaped internment and having safely returned to England joined Drake Battalion which in February 1915 was sent out to Gallipoli. He survived the worst of the campaign but was severely wounded in the chest by a dumdum bullet losing a lung on the last day of the Gallipoli evacuation.

He subsequently joined the *Royal Sovereign* and became the officer in charge of 'A' turret and the Forecastle Division. Later he joined the staff of Admiral Leveson in the *Barham*, where he spent the rest of the war until demobilised. When the *Barham* and her squadron were the guardships for the German ships interned at Scapa Flow, Lieut.-Commander Wells was one of the Staff Officers responsible for security, patrolling the anchorage in a drifter.

During the Second World War, Wells was recalled to duty in 1940 and became the Training Commander at the Royal Naval Air Station at Lee-on-Solent and in 1943 became President of one of the Naval Air Command Interview Boards.

For a time in the 1920's Wells was Secretary of the Union Club when it moved from Trafalgar Square to Carlton House Terrace. He was a good cricketer, being secretary of the Incognito Cricket Club and once played for an England XI against the West Indies at Lords.

He died in 1946 at Felixstowe while on his end of war leave.

Captain John Rustat Hemsted CB VD RNVR

Captain J. R. Hemsted

Jack Hemsted, known to the sailors as 'Hemel' joined the Division on the 24th November 1903 at Fishmongers' Hall and was detailed to No. 4 Company. He became a signalman and his keenness for the R.N.V.R. was demonstrated by his consistently high attendances. Invariably his name was at the top of the list. Also he faithfully kept scrapbooks which London Division is now lucky enough to possess and which have been of immense value in the research for this book.

Returning from the First War he was commissioned in 1921 as a Lieutenant.

It took two tragedies – the drowning of Commodore King and the ill health of Captain Wells – to bring Hemsted to the helm of the Division he so clearly loved. As a Lieut.-Commander in 1932 he retired and was given the rank of Commander. A year later he was called back to take command and was promptly promoted to Captain. In civilian life he was a director of Courage Brewery and is reputed to have found jobs as draymen for those in the Division who were unemployed in the difficult years of the Depression. When recruiting a new young Chaplain in 1937, Captain Hemsted said to him 'If you join you will not be expected to go about with the open Bible in your hand, but you will not be forgiven if it is far from your heart' – words which Bishop George Reindorp has not forgotten in forty-five years.

Hemsted remained in command of the Division until after the outbreak of the Second World War, when the Division ceased to function once again, but on 13 November 1939 he was appointed to various Admiralty Selection

Boards, serving eventually in H.M.S. *King Alfred* at Hove, interviewing for officer selection. Also serving at Hove at this time in the same capacity was the former executive officer of London Division, Captain D. B. Nicholson. Hemsted finally retired from such duty, on age, in 1942 at the age of sixty, thus ending forty remarkable years of devoted service. He died in 1953 and the senior chaplain of London Division, then The Reverend Gordon Taylor, gave the address at his funeral service in Southwark Cathedral.

The Battle of Jutland

Chief Yeoman of Signals J. R. Hemsted

A personal account of the naval engagement on 31 May 1916 by Chief Yeoman of Signals J. R. HEMSTED on board the *Iron Duke*.

The Battle Fleet were standing in a south easterly direction on the 31st in the usual formation – 6 divisions in line abreast with the cruisers screening ahead, and Destroyers acting as submarine screen just ahead of the divisions. The first real intimation we had was the signal 'Prepare for immediate action' hoisted in the early afternoon. Immediately the hands were piped to tea (it being then about 3.10 pm) and Quarters was sounded at 4.0 pm. A signal was made stating that our Battle Cruisers were engaging the German Battle Cruisers and afterwards the German High Seas Fleet which were being driven in a Northerly direction which was toward our Fleet. We heard afterwards that the four Dreadnoughts of the Queen Elizabeth class namely *Barham* (Rear Admiral), *Malaya, Valiant* and *Warspite* (the *Queen Elizabeth* was refitting), were with the Battle Cruisers under Vice Admiral Beatty.

We first heard heavy firing about 70 green which we supposed to be some distance away and blown in our direction by a breeze from the Southward, but shortly afterwards the Battle Cruiser Fleet were made out steaming across our bows firing well on an extreme forward bearing to Starboard. Just before our Battle Cruisers were sighted our Cruiser-screen (stationed about 10 miles ahead) were seen to be firing at something we could not see, as the weather was very misty. The Battle Cruisers were *Lion, Princess Royal, Queen Mary, Tiger* and *New Zealand,* and shortly afterwards the 5th Battle Squadron came obliquely across toward the Battle Fleet from the Starboard Wing. The Battle Fleet altered course in succession to starboard, about three points, then deployed to port into line of battle, the Battle Cruisers being ahead and the 5th Battle Squadron on the rear end of the line. Just at this time we sighted a large amount of steam on the starboard beam which was made out to come from a three funnelled German Cruiser badly damaged and sinking. The Battle Cruisers on passing put a few more salvoes into her and the *Iron Duke,* to clear the bores of her guns, fired two salvoes but no hits were observed. All other ships within range were potting at her especially a Cruiser of the Shannon class which got in front of our Battle line, made a terrific quantity of smoke and obscured everything on the firing side. She eventually drew off toward head of line. We then sighted 3 or 4 German battle ships of the 'Konig' class which appeared in the mist about 68 green. The error given by the range finding plot was 500 yards, the actual range being 11,000 yards. We then opened fire on one of them it being 6.26 pm. The first salvo fell short, the second straddled, the third – one hit, and, the fourth – three hits, all before the foremast. She then altered course about twelve points to starboard and was lost in the mist. Two more salvoes were fired and appeared to fall short. We checked fire and saw an object in the water on the starboard bow which was believed to be a zeppelin but was a ship cut in two amidships with only her bows and stern showing vertically. The destroyer *Badger* was picking up survivors of which there were dozens swimming and clinging to the sunken ship, and there were numerous mangled bodies in the water. On passing and asking *Badger* who the ship was, she replied *Invincible* (Flagship of Rear Admiral Hood). Shortly afterwards the mist lifted and we sighted what appeared to be a battle ship, very much like our Queen Elizabeth, but it was evidently an enemy ship, as several of our own ships were firing at it. We therefore opened fire at it at a range of 16,000 yards but the inclination was judged to be *toward* instead of *away* from us and our shots fell a long way to the right and by the time the deflection was corrected and in all 4 salvoes were fired, she was out of sight as the mist lowered again. When last seen she apparently had a number of German Destroyers around her. Fire was checked and shortly afterwards another Battleship came in sight but before fire could be opened she was obscured by smoke. Her bearing was 93 green. Nothing happened for some time after this, but eventually a Destroyer attack developed against us. We opened fire with our 6 inch guns. These fired continuously for about 10 or 15 minutes but the results were not seen. The turrets were then directed on to the destroyer and one salvo was fired and the Director Gunner is convinced that we hit her. After smoke lifted the target was nowhere to be seen. No firing was then carried on by us. This was the last occasion of engaging the enemy. Heavy firing was observed on the Starboard bow and the Light Cruiser, *Comus* was seen to be hit amidships and a flash lit up the whole of the main deck. However she continued to proceed. When it got dark, course was altered to the south and we appeared to have got well ahead of the High Sea Fleet. During the night firing was observed at intervals right away to the Starboard Quarter, and huge red glares were seen, and a few star shells, at long intervals. (In the morning this turned out to be a successful attack carried out by our Destroyers). We appeared to be in a very favourable position for a Fleet action in the morning. Of course a strict look-out was kept all night, all hands closed up at action stations. At 2.0 am Quarters was sounded again. Just before 3.0 am a Zeppelin was sighted coming astern. Her range (which was very difficult to get) was about 20,000 to 21,000 yards which was well out of range. Some ships opened fire on her for a few minutes but *Iron Duke* did not as it was a waste of ammunition. She then disappeared and, I suppose, piloted the enemy into harbour as they were seen no more. *Iron Duke* fired in all 18 rounds per turret.

Marlborough was hit by torpedo during the action but continued to proceed at 17 knots and left the Fleet in the evening for her base. The Vice Admiral, Second-in-Command shifted his Flag to *Revenge*. *Defence* also blew up on our Starboard Quarter in the early part of the action. She was the Flagship of Rear Admiral Arbuthnot of the First Cruiser Squadron.

Many other incidents and brave deeds all happened of course, as in all engagements, which were not seen, and more which it would take too long to recount, but next time we mean to teach them the biggest lesson they've had.

GOD SAVE THE KING.

Commodore Charles Patrick Cay Noble, CBE DSC VRD DL RNR

Commodore Charles Noble

In some respects Charles Noble's R.N.V.R. career was not unlike those of Commodore King and Captain Wells. All of them joined the Division as junior officers in the days of peace prior to a world war; all of them served with distinction in war, returning to assist in the re-establishment of the Division in post war years, and all three eventually assumed command.

Noble was, however, more fortunate than King and Wells in not being required to fight on land. When war broke out in 1939 he went to the Fleet in the appointment for which he had been trained, as the Lieutenant in command of one of the four Anti-Aircraft units, and to the life at sea he would have envisaged when he joined London Division as a midshipman in 1932.

A gunnery specialist, he was a member of the R.N.V.R. long course at Whale Island in 1941 and he became the second of a trio of gunnery officers

Fishmongers' bowl

to command London Division in the Fifties. In 1958 he was appointed the first Commodore (List 3) of the new Royal Naval Reserve.

It is to Commodore Noble's initiative that London Division owes its renewed, and now continuing, link with the Fishmongers' Company of which he is a past Prime Warden. In 1961, during the Prime Wardenship of His Royal Highness The Duke of Edinburgh, London Division was privileged to be adopted by the Fishmongers' Company and, at a dinner on board the *President* to mark the occasion, the Company presented to the Wardroom a silver bowl in the shape of a fishmonger's hat. This always has pride of place on the top table at a Wardroom dinner.

Commodore Noble was a stockbroker in the City by profession and he shares with Captain Peter Rees the distinction of being the only commanding officers to be Deputy Lieutenants of both the City of London and of Greater London. He was High Sheriff of Greater London in 1977.

Commodore James Young

Commodore James Goddard Young, CBE DSC VRD DL RNR

In 1958 it might have seemed to an onlooker that to be chosen to command London Division it was necessary to be a gunnery officer with a commanding presence, a charming manner and a strong voice, preferably with a talent for piano playing and acting. James Young had all these attributes but it was his professionalism, dedication and enthusiasm which made him a natural choice for the prize of command. He was the third of the trio of such gunnery officers who became Captain and shares with his immediate predecessor, Charles Noble, musical and acting ability. Many will recall with pleasure their solo and duet pieces on the piano.

As a chartered accountant in the firm of Andrew Weir, of which he eventually became Finance Director, it was not surprising that Young entered the Division in 1935 as a Paymaster Sub Lieutenant. But always a man of action, he transferred to the Executive Branch specialising in gunnery. After his distinguished war service he returned to London Division and made his way steadily to the top, succeeding Noble in 1961 as Commodore of the R.N.R.

The courage and sense of duty which Young so amply displayed in his R.N.V.R. career was demonstrated again in 1970 when he tackled an armed robber in the City for which he was given a Binney Award for gallantry. The Division was saddened by his sudden death at Christmas 1982 and is grateful for the valuable contribution he made to the history.

D.M.M.

3

Commanding Officers

Captain Viscount Elveden, CB CMG VD ADC (formerly Captain The Hon. Rupert E.C.L. Guinness)	1903–1920
Commodore H.D. King, CB CBE DSO VD	1920–1930
Captain N.ff. Wells, VD	1930–1933
Captain J.R. Hemsted, CB VD	1933–1939
Captain R.T. Janson, VRD	1946–1949
Captain C.B. Sanders, CBE VRD	1949–1952
Captain J.A. Creed, VRD	1953–1955
*Captain (later Commodore) C.P.C. Noble, DSC VRD	1955–1958
*Captain (later Commodore) J.G. Young, DSC VRD	1958–1961
Captain I.A.B. Quarrie, CBE VRD DL	1961–1964
Captain J.B. Leworthy, VRD ADC	1964–1967
Captain A.H. Spratt-Kerswill, VRD ADC	1967–1970
Captain B.J. West, VRD ADC	1970–1973
Captain R.St.J. Fancourt, RD	1973–1976
Commander A.P.M. Woodward, VRD	1974
Captain P.S. Rees, RD ADC DL	1976–1979
Captain G.K. Beattie, RD ADC DL	1979–

** later appointed Commodore (List 3)*

Executive Officers (Commanders)

Lieutenant E. Wildy (Second-in-Command)	1903–1914
Commander E. Wildy, OBE VD	1921
Commander N.ff. Wells, VD	1922–1930
Commander L.L. Withrington, VD	1931–1938
Commander B.D. Nicholson, VD	1938–1939
Commander C.B. Sanders, VRD	1946–1948
Commander I. MacGregor, VRD	1949
Commander J.A. Creed, VRD	1950–1952
Commander C.P.C. Noble, DSC VRD	1952–1954
Commander J.G. Young, DSC VRD	1955–1957
Commander I.A.B. Quarrie, VRD	1958–1960
Commander J.B. Leworthy, VRD	1961–1963
Commander J.B. Currie, VRD	1964
Commander A.H. Spratt-Kerswill, VRD	1964–1966
Commander R.R. Gordon, OBE VRD	1966–1968
Commander B.J. West, VRD	1968–1969
Commander J.S. Wordie, VRD	1970–1972
Commander R.St.J. Fancourt, RD	1972–1973
Commander A.P.M. Woodward, VRD	1973–1975
Commander P.S. Rees, RD	1975–1976
Commander G.K.Beattie, RD	1976–1978
Commander J.G.N.T. Cosnett, OBE RD	1979–1981
Commander D.W. Green, RD	1981–1982
Commander H.R. Roberts, RD	1982–

Heads of Department

Senior Supply Officers
From 1947
*Lieutenant Commander F.A. Kemmis Betty
Commander A. Chappell
*Commander (later Captain) F.A. Kemmis Betty OBE VRD
Commander R. Borner OBE VRD
Commander P.J. Douglas VRD
Commander R. Wanless VRD
*Commander (later Captain) W.J. Parsons VRD
*Commander (later Captain) A.B. Sainsbury VRD
Commander A.J. Phillips VRD
Commander J.A. Barnett RD
*Commander B.C. Macleod RD
Commander B.M. Cole RD
Commander C.P. Powlett RD
**Later appointed Senior Supply Officer R.N.R.*

Senior Engineer Officers
From 1947
*Commander (later Captain) W.G. Smith VRD
Commander J.M. Hargreave VRD
Commander R.W.E. Martin VRD
Commander P. Barlow VRD
Commander G.G. Davies VRD
Commander The Hon. E.B. Evans VRD
Commander E.L. Eavis, VRD
Commander W.J.G. Smith RD
Lieutenant Commander M. Bedwell RD
**later appointed Senior Engineer Officer R.N.R.*

Senior Electrical/Weapons Electrical Officers
From 1948
Lieutenant Commander R.E. Davis VRD 1948–1951
Commander R.C.R. Brooke VRD 1951–1962
Commander D.A. Barnes VRD 1962
Commander R. Puttock VRD 1962
Commander S. Beveridge VRD 1962–1965
*Commander (later Captain) J.E. Tolkein VRD 1965–1969
*Commander (later Captain) L.W. Smith RD 1969–1972
Commander J.E. Smith RD 1972–1975
*Commander S.H. Cooper RD 1975–1979
Lieutenant Commander A. Lane RD 1979 –

Senior Technical Officer
From 1981
Commander A.H. Stephenson RD 1981–
**later appointed Senior Weapons Electrical Officer R.N.R.*

Senior Naval Control of Shipping Officers
From 1973
Commander P.H. Ramsden OBE RD 1973–1977
Commander N.W. Hodges RD 1977–1980
Commander T.C. Haile RD 1980–

Principal Medical Officers
From 1920
Surgeon Captain A.R. Brailey VD MC 1920–1930
Surgeon Captain J.B. Ronaldson OBE VD KHP 1930–1939
Surgeon Captain G.F. Abercrombie VRD KHP 1947–1951
Surgeon Captain H.M. Willoughby VRD QHP 1951–1956
Surgeon Captain P.de B. Turtle VRD QHP 1956–1958
Surgeon Commander (later Captain) A.D. Petro VRD QHP 1958–1959
*Surgeon Commander (later Captain) A.W. Hagger VRD QHP 1959–1963
*Surgeon Commander (later Captain) E.C. Glover VRD QHS 1963–1966
Surgeon Commander H.R. Gray VRD 1966–1967
*Surgeon Commander (later Captain) F.S. Preston VRD QHP 1967–1970
*Surgeon Commander (later Captain) A.P. Davidson RD QHP 1970–1974
*Surgeon Commander (later Captain) D. Churchill-Davidson RD QHP 1974–1977

Surgeon Commander A.N. Hepburn RD 1977–1980
Surgeon Commander R.J. Berry 1980–
**later appointed Senior Medical Officer, R.N.R.*

Senior Dental Surgeons
From 1947
*Surgeon Captain (D) R. Symmons VRD 1947–1958
Surgeon Commander (D) E.V.B. Widdowson VRD 1959
*Surgeon Commander (D) (later Captain) J. Symmons VRD QHDS 1959–1960
*Surgeon Commander (D) (later Captain) B.W. Pett VRD QHDS 1960–1963
Surgeon Commander (D) R.E. Matthews VRD 1963–1967
*Surgeon Commander (D) (later Captain) M.N. Naylor RD QHDS 1968–1973
*Surgeon Commander (D) (later Captain) H.K. Kemp RD QHDS 1974–1977
Surgeon Commander (D) H. Cannell RD 1977–
**later appointed Senior Dental Officer R.N.R.*

Senior Chaplains
From 1920
The Reverend A. Stainsby
The Reverend O.R. Fulljames VRD
The Reverend G.C. Taylor VRD
The Reverend K.N.J. Loveless VRD
The Reverend G.C. Cutcher VRD
The Reverend M. Shrewsbury
The Reverend P.E.H. Golding RD
The Reverend B.H. Talbott RD

Roman Catholic Chaplains
From 1947
The Reverend W. Meyjes
The Reverend B.C. Wilkins
The Reverend J. Armitage

Free Church Chaplains
The Reverend E.A. Way

Sub Lieutenants of the Gunroom
From 1957
P. Garner 1957–58
C.D.G. Roberts 1958–59
H.R. Roberts 1959–60
M. Greenslade 1960–61
W. Cross 1961–62
R.J. Tucker 1962–63
J.C. Gunner 1963
P.J. Crean 1963–64
P.S. Albertini 1964-66
R.M.C. Topham 1966–67
D.A. Grant 1967–68
I.S. Howat 1968–69
W. Richmond 1969–70
M.E. Hayton 1970–71
R.S. Fidgen 1971–72
H.J. Nell 1972–73
C.M.E. Riley 1973–74
R.M. Thornborough 1974–75
M.T. Bostock 1975–76
J.R.A. Hanratty 1976–77
R.G. Avis 1977–78
M.A. Savory 1978–79
A. Halsey 1979–80
C.K.C. Ord 1980–82
J.N. Donne 1982–

WRNR Unit Officers
From 1951
Chief Officer O. Snow 1951–1955
Chief Officer M.H. Love MBE 1955–1958
Chief Officer B. Meeke VRD 1958–1961
Chief Officer H.L. Berger VRD 1961–1964
Chief Officer E.M. Copley VRD 1964–1967
First Officer E.W. McCreath RD 1967–1970
*First Officer (later Chief Officer) D.M. Mason RD 1970–1973
First Officer A.M. Goodhart RD 1973–1975
**Later appointed Senior Officer W.R.N.R.*

1975 the W.R.N.R. ceased to be administered as a separate unit

Presidents of Chief and Petty Officers Mess

G.L. McDonell C.P.O. 1904–1906
J. Crowley C.P.O. 1907
W. Clegg Smith C.P.O. 1908
H.A. Lockett C.P.O. 1908
C.E. Jupp C.P.O. 1910
A.J. Evans C.P.O. 1911
W.E. Menday C.P.O. 1912
C.G. Murray C.P.O. 1913
J.R. Hemsted Yeo. Sigs. 1914
A.C. Forman C.P.O. 1922–1923
F.W. Innes Ch. Wtr. 1924
J.P. Clemens C.P.O. 1925
J.A. Budd C.P.O. 1926
G.F. Ashurst C.P.O. 1927
A.J. Creedy Ch. Yeo. Sigs. 1928
H.E. Edwards E.R.A. 1929
J.W. Wiggins C.P.O. 1930
J.A. Budd C.P.O. 1931
R.V. Redfearn E.A. 1932
J. Bunker C.P.O. 1933–1935
T.R. Parsons C.P.O. Tel. 1936
W. Carter Ch. Yeo. Sigs. 1937
J.E. Bryon Ch. Wtr. 1938
A.E. Ingram C.P.O. 1939
R. Mason C.P.O. 1948–1951
R. Mason C.P.O. }
A. Hunt E.R.A. } 1952
E.F. Perkins C.P.O. 1953
E.J. Wakely C.P.O. 1954
F. Fitches C.P.O. 1955–1956
R.E. Minter C.P.O. 1957
J.E. Hunt E.R.A. 1958–1959
M.J. Talbot E.A. 1960
P.E. Cullum C.R.E. 1961
J.J. McNaughton Ch. Jnr. 1962
E.G. Glover E.R.A. 1963
S.T. Jefferson DSM C.P.O. 1964
L.H. Barker Ch. M(E) 1965
B.T. Lister Ch. Coder 1966
J.E. Hunt BEM, C.P.O. 1967
A.F. Reed C.P.O. 1968
C.R. Turner C.P.O. 1969–1970
D.F. Creasy C.P.O. 1971–1972
R.J. Trodd C.R.S. 1973–1974
R. Spears C.P.O. 1975
R. Cotton C.C.Y. 1976–1978
J. Batchelor R.E.A. 1979
M.W. Bird C.M.E.M.(M) 1980–1982
R.V. Abbott BEM, C.M.E.M.(M) 1982–

Commander-Instructors/Chief Staff Officers 1903–1980

Commander W. Hewetson R.N.(Rtd.) 1903–1908
Lieutenant R.M. Tabutson R.N.(Rtd.) 1908–1910
Lieutenant N.C.A. Moore R.N. 1910–1912
Lieutenant R.G. Talbot R.N. 1912–1914
Lieutenant J.P. Gibbs R.N. 1914

Commander D.W. O'B. Forsyth R.N. 1920–1922
Lieut. Commander J.H. Thom OBE R.N. 1922–1924
Lieut. Commander H.H.J.F. Teale R.N. 1924–1926
Lieut. Commander K. Michell MVO DSC R.N. 1926–1928
Lieut. Commander A.M.C. Stileman R.N. 1928–1930
Lieut. Commander W.F. Smithwick DSC R.N. 1930–1933
Lieut. Commander E.R. Lewis DSC R.N. 1933–1935
Lieut. Commander G.C. Harris R.N. 1935–1939
Lieut. Commander H.C. Skinner OBE, R.N. 1939

Commander E.R. Lewis DSO DSC R.N.(Rtd.) 1947–1948
Commander J.G. Gould OBE R.N. 1948–1949
Lieut. Commander A.E. Doran DSC R.N. 1949–1951
Commander C.G. Forsberg R.N. 1951–1952
Commander M.B.St. John DSC R.N. 1952–1955
Commander C.J. Bateman DSC R.N. 1955–1956
Commander D.R.E. Calf DSC R.N. 1956–1958
Commander J. Marriott R.N. 1958–1959
+ Commander O.N.A. Cecil R.N. 1959–1961

Commander A.A. Davies R.N. 1961–1963
*Commander J.M.H. Cox R.N. 1963–1964
Commander J.R.B. Montanaro R.N. 1964–1966
Commander E.J. Sawkins R.N. 1966–1968
Commander A.G.F. Crosbie R.N. 1968–1970
Commander E.V.F. Savill R.N. 1970–1973
Commander R.W. Palastre MBE R.N. 1973–1976
Commander J.G.F. Slocock R.N. 1976–1978
Commander J.P. Marriott R.N. 1978–1980
Commander G.W.G. Hunt R.N. 1980

+Rear Admiral Sir Nigel Cecil KBE CB currently Governor of the Isle of Man
**Vice Admiral Sir John Cox KCB currently Flag Officer Naval Air Command*

W.R.N.S. Staff Officers

Second Officer M. Hartridge 1951–1952
Second Officer S. Woodcock 1952–1954
Second Officer P. Field 1954–1955
Second Officer M.G. Cheyney 1955–1957
*Second Officer S.V.A. McBride 1957
Second Officer M.A. Brown 1957–1959
Second Officer M.A. Wydham Thomas 1959–1960
Second Officer S.M. Audrey 1960–1962
Second Officer N.L. Bugden 1962–1964
Second Officer P.M. Williams 1964–1966
Second Officer L.F. Moody 1966–1967
Second Officer R. Pitts 1967–1968

**Commandant S.V.A. McBride CB Hon.ADC*
Director, Women's Royal Naval Service 1976–1979

Admiral Commanding Coast Guard & Reserves 1903–1923
Admiral Commanding Reserves from 1923–1977
Commanders-in-Chief Naval Home Command 1977–1983

Admiral Ernest Rice 1903–1905
Vice Admiral Reginald F.H. Henderson, CB 1905–1909
Vice Admiral Frederick S. Inglefield, KCB 1909–1913
Vice Admiral Sir Arthur M. Farquhar, KCB CVO 1913–1915
Vice Admiral The Hon. Sir Alexander E. Bethall, KCB KCMG 1915–1916
Vice Admiral The Hon. Sir Somerset A. Gough-Calthorpe, KCB CVO 1916–1917
Vice Admiral Sir Cecil F. Thursby, KCMG 1917–1918
Vice Admiral Sir Dudley R.S. De Chair, KCB MVO 1918–1921
Vice Admiral Sir Morgan Singer, KCVO CB 1921–1923
Vice Admiral Sir Hugh H.D. Tothill, KCB KCMG KCVO 1923–1925
Vice Admiral Sir Lewis Clinton-Baker, KCB KCVO CBE 1925–1927
Vice Admiral Sir Arthur A.M. Duff, KCB 1927–1929
Vice Admiral Sir John D. Kelly, KCB 1929–1931

Vice Admiral H.W. Parker, CB CMG 1931–1933
Vice Admiral Sir George K. Chetwode, KCB CBE 1933–1936
Vice Admiral Sir H.J.S. Brownrigg, KBE CB DSO 1936–1938
Vice Admiral Sir Noel F. Laurence, KCB DSO 1938–1941
Rear Admiral V.S. Butler, DSO 1941
Vice Admiral J.G.P. Vivian, CB 1941–1945
Vice Admiral C.E. Morgan, CB DSO 1945–1947
Vice Admiral Sir Wilfred R. Patterson, KCB CVO CBE 1947–1949
Vice Admiral The Hon Guy H.E. Russell, CB CBE DSO 1949–1950
Vice Admiral W.R. Slayter, CB DSO DSC 1950–1952
Vice Admiral J.A.S. Eccles, CB CBE 1952–1953
Rear Admiral A.K. Scott-Moncrieff, CB CBE DSO 1953–1955
Vice Admiral J.W. Cuthbert, CB CBE 1955–1956
Rear Admiral G. Thistleton-Smith, CB GM 1956–1958
Vice Admiral W.K. Edden, CB OBE 1958–1960
Vice Admiral R.A. Ewing, CB DSC 1960–1962
Rear Admiral H.C. Martell, CBE 1962–1965
Rear Admiral G.H. Carew-Hunt 1965–1968
Rear Admiral B.C.G. Place, VC DSC 1968–1970
Rear Admiral I.D. McLaughlan, CB DSC 1970–1972
Rear Admiral I.G.W. Robertson, DSC 1972–1974
Rear Admiral H.W.E. Hollins, CB 1974–1977

Admiral Sir Terence Lewin, GCB MVO DSC ADC 1977
Admiral Sir David Williams, GCB ADC 1977–1979
Admiral Sir Richard Clayton, GCB ADC 1979–1981
Admiral Sir James Eberle, GCB ADC 1981–1982
Admiral Sir Desmond Cassidi, GCB ADC 1982–

NOTE From 1955 to 1977 Admiral Commanding Reserves flew his flag in HMS *Discovery*.

London Division Sea Tenders

1948–1983

HMS Thames	Fast Despatch Boat	FDB 76	1949–1951
HMS Thames	Motor Minesweeper	MMS 1789	1952–1954
HMS Thames (ex Alverton)	Coastal Minesweeper	M 1104	1954–1961
HMS Thames (ex Buttington/ex Venturer)	Coastal Minesweeper	M 1117	1961–1967
HMS Dufton	Coastal Minesweeper	M 1145	1967–1969
HMS Thames (ex Woolaston)	Coastal Minesweeper	M 1194	1969–1975
HMS Sandpiper	Patrol Vessel	P 263	1977–1979
HMS Glasserton	Coastal Minesweeper	M 1141	1979–1980
HMS Lewiston	Coastal Minesweeper	M 1208	1981–
HMS Isis	Fast Despatch Boat	FDB 80	1948–1949
HMS Isis	Motor Minesweeper	MMS 1785	1950–1954
HMS Isis (ex Pulham)	Inshore Minesweeper	M 2721	1955–1963
HMS Isis (ex Cradley)	Inshore Minesweeper	M 2010	1963–1973

An artist's impression by Commander G. W. G. Hunt R.N. of the new generation of Fleet Minesweepers, ordered in 1982.

ROLL OF HONOUR

Royal Naval Division Memorial — Greenwich

The following is the inscription drawn up by Captain E.G. Roper, DSO, DSC, R.N., on behalf of the Admiral President of the Royal Naval College, Greenwich, which has been framed and attached to the Divisional Memorial.

The Royal Naval Division

The Royal Naval Division was formed in August, 1914, by the personal direction of Mr. Winston Churchill, then First Lord of the Admiralty, from Royal Marines and Naval Reservists (R.F.R., R.N.R., and R.N.V.R.) not immediately required to man the Fleet.

The Division fought throughout the war of 1914—1918, at Antwerp, on Gallipoli, and Salonika and finally from 1916 to 1918 in France and Belgium and suffered the following casualties:—

KILLED: 582 Officers 10,797 other ranks.

WOUNDED: 1,364 Officers 29,528 other ranks.

Owing to the fearful losses suffered in November, 1916, several Army Battalions were attached to the Division from time to time.

The following decorations were earned:—

VC	8
DSO	42
DSC	17
MC	137
DCM	53
CGM	23
DSM	79
MM	555
MSM	9

These were the men of whom Sir Winston Churchill wrote:—

"By their conduct in the forefront of the battle, by their character, and by the feats of arms which they performed, they raised themselves into that glorious company of the seven or eight most famous Divisions of the British Army in the Great War. Their reputation was consistently maintained in spite of losses of so awful a character as to sweep away three or four times over the original personnel. Their memory is established in history and their contribution will be identified and recognized a hundred years hence from among the enormous crowd of splendid efforts which were forthcoming in this terrible period. Deriving as they did their nomenclature, their ceremonial, their traditions, their inspiration from the Royal Navy, they in their turn cast back a new lustre on that mighty parent body of which it will ever be proud and for which it must ever be grateful.

"Long may the record of their achievements be preserved, and long may their memory be respected by those for whom they fought."

London Division Roll of Honour

1914–1918

No. 1 Company
A.S. Downes
F.B. Ramsey
F.B. Bennett
J.A. McMillan
A.P. Nasmith
B.H. Ellis
C.H. Tamplin
R.A. Battson
H.M.S. Carpenter
D.G. Smith
H.W. Troughton
L.H. Redding
H.A. Pailthorpe
H.E.A. Sparks
A.C. Davis
A.V. Hunt
A.G.L. Taylor
B.S. Clemens
V. Hollinton
C.E.C. Flood
W.E. Dyke
G.P. Cooke
C.A. Southin
G.M. Davis
G.W. Telfer
G.W. Ross
G.N. Strang
W.G.E. Phillips
L.J. Seeds
J.W. Edwards
E. Manners Ridge

No. 2 Company
J.F. Bunce
E.J. Palmer
D. Robertson
A. Bazard
W.J. Scott
H.J. Hodge
G.W. Hodge
L.J. Pincombe
H.J. Selkirk
C.J. Kerswell
A.H.T. Vizard
R.S. Rogers
R.J. Gee
F.R.G. Sharp
W.C. Jameson
C.E. Ireland

No. 3 Company
F. Newland
O.C. Beale
R. Brouse
S.D. Borland
G. Bartlett
A.F. Duprey
A.E.N. Chance
E. Fluit
E.L. Franklin
C.E. Withey
C.A. Morris
A.S. Ashford
H.W. Radcliffe

No. 4 Company
W.E. Barry
H.A. Massey
A. Brown
H.W. Little
C.F. Lan-Davis
E.E.J.A. Behn
H.D. Allen
J.A. Fairlie
R. Page
G.H. Breeden
A. Donne
W.V. Murphy
C.P. Asbury
L.H. Watts
A.R. Nash
R.H. Guppy
W. Manning
F.W.B. Gill
K.O. Owen
L.E. Whitehead
D.J. Crowley
C.S. Carter
R.J.S.W. Douglas
J.C.S. Warwick

No. 5 Company
F.A. Lowe
H.M. Cook
P. Hagges
D.R.G.P. Alldridge
P.A. Rootham
R.G. Hayes
H.G. Hayward
C.D. Marshall

No. 6 Company
A.E. Barter
F.C. Baxter
J. Beech
E.C. Brooks
G.D. Doucet
R.J. Jordan
E.F. Loveless

No. 7 Company
O.H. Hanson
J.F. Binnington
L.G. Black
T.C. Brightwell
H.E.F. Buch
J.W. Dowling
J.D. Ewings
P.C. Garnham
H.J. Gore
R.H. Guthrie
P.E. Hedger
F.A. Meggy
G.H. Millar
H. Raine
E.T. Saunders
B.J. Sercombe
R.J. Smith
R.J.S. Spain
E.W. Tice
W.S. Wheeler

No. 8 Company
T.F. Adams
C. Baker
E.J. Bowles
L.A. Cherry
S.H. Cripps
C. Durnham
E.C. Harris
E.H. Imison
T.E. Love
T. McCreath
L.C. Richardson
R. Stevens
H.E. Walton
W.R. Williamson

No. 9 Company
G.R. Airey
W.H. Anderson
W.H. Baker
F. Benn
A.W. Darvill
C. Dover
A.C. Hawkins
J.S. Henderson
W.A. Isaacs
S.H.H. Ixer
R.W. King
R.R. Lawrence
J.R. McCormick
A. Mattinson
B.W. Smyth
G.J. Stavert
W. Tompsett
F.L. Tutt
R.J. White
R.R.C. Winter

No. 10 Company
E.A. Armstrong
L.W. Banks
M.S. Barrand
K.F.H. Bell
F.E. Bowers
J.W. Campbell
W.R. Chandler
F.C. Gardener
F.R.W.G. Hall
S.F. Harris
J.W. Lane
F.G. Liddiard
W.J. McLellan
G. Paterson
W.E. Rainbow
R.W.C. Shaw
T. Staples
E.S. Stephens
P.A. Stiff
A.W.St.C. Tisdall
E. Tolra
S.C. Wakefield
S.F. Webb
E.R. Wilkins
K.O. Williams
H. Withers

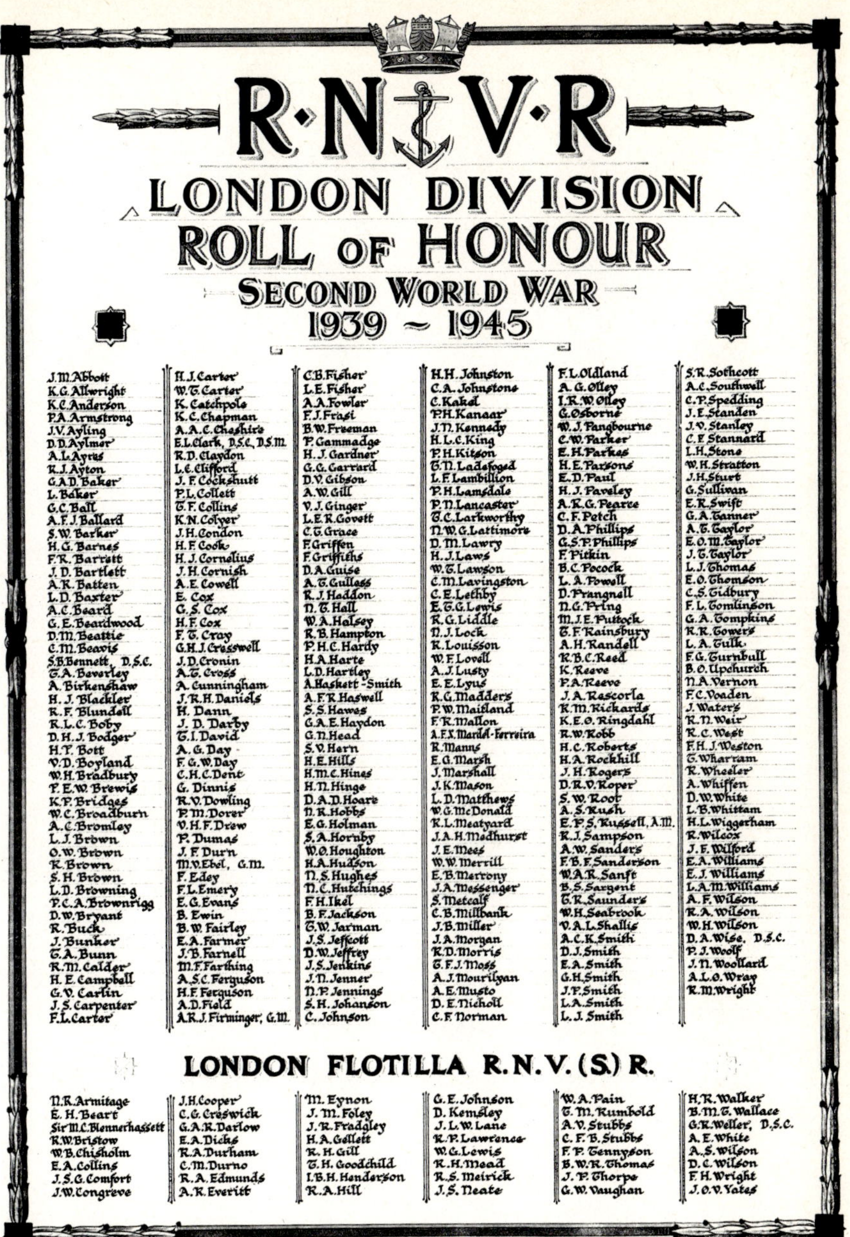

R·N·V·R

LONDON DIVISION

ROLL OF HONOUR

SECOND WORLD WAR

1939 ~ 1945

J.M. Abbott
K.G. Allwright
K.C. Anderson
P.A. Armstrong
J.V. Ayling
D.D. Aylmer
A.L. Ayres
R.J. Ayton
G.A.D. Baker
L. Baker
G.C. Ball
A.F.J. Ballard
S.W. Barker
H.G. Barnes
F.R. Barratt
J.D. Bartlett
A.R. Batten
L.D. Baxter
A.C. Beard
G.E. Beardwood
D.M. Beattie
C.M. Beavis
S.B. Bennett, D.S.C.
T.A. Beverley
A. Birkenshaw
H.J. Blackler
R.F. Blundell
R.L.C. Boby
D.H.J. Bodger
H.T. Bott
V.D. Boyland
W.H. Bradbury
T.E.W. Brewis
K.P. Bridges
W.C. Broadburn
A.C. Bromley
L.J. Brown
O.W. Brown
R. Brown
S.H. Brown
L.D. Browning
P.C.A. Brownrigg
D.W. Bryant
R. Buck
J. Bunker
T.A. Bunn
R.M. Calder
H.E. Campbell
G.V. Carlin
J.S. Carpenter
F.L. Carter

N.J. Carter
W.T. Carter
K. Catchpole
K.C. Chapman
A.A.C. Cheshire
E.L. Clark, D.S.C., D.S.M.
R.D. Claydon
L.C. Clifford
J.F. Cockshutt
P.L. Collett
T.F. Collins
K.N. Colyer
J.H. Condon
H.F. Cook
H.J. Cornelius
J.H. Cornish
A.E. Cowell
E. Cox
G.S. Cox
H.F. Cox
F.T. Cray
G.H.J. Cresswell
J.D. Cronin
A.T. Cross
A. Cunningham
J.R.H. Daniels
H. Dann
J.D. Darby
T.I. David
A.G. Day
F.G.W. Day
C.H.C. Dent
G. Dinnis
R.V. Dowling
P.M. Dorer
V.H.F. Drew
P. Dumas
J.F. Durn
M.V. Ebel, G.M.
F. Edey
F.L. Emery
E.G. Evans
B. Ewin
B.W. Fairley
E.A. Farmer
J.B. Farnell
M.F. Farthing
A.S.C. Ferguson
H.F. Ferguson
A.D. Field
A.R.J. Firminger, G.M.

C.B. Fisher
L.E. Fisher
A.A. Fowler
F.J. Frasi
B.W. Freeman
P. Gammadge
H.J. Gardner
G.G. Garrard
D.V. Gibson
A.W. Gill
V.J. Ginger
L.E.R. Govett
C.T. Grace
F. Griffen
F. Griffiths
D.A. Guise
A.T. Gulless
R.J. Haddon
N.T. Hall
W.A. Halsey
R.B. Hampton
P.H.C. Hardy
H.A. Harte
L.D. Hartley
A. Haskett-Smith
A.F.R. Haswell
S.S. Hawes
G.A.E. Haydon
G.N. Head
S.V. Hern
H.E. Hills
H.M.C. Hines
H.N. Hinge
D.A.D. Hoare
N.R. Hobbs
E.G. Holman
S.A. Hornby
W.O. Houghton
H.A. Hudson
N.S. Hughes
N.C. Hutchings
F.H. Ikel
B.F. Jackson
T.W. Jarman
J.S. Jeffcott
D.W. Jeffrey
J.S. Jenkins
J.N. Jenner
N.P. Jennings
S.H. Johanson
C. Johnson

H.H. Johnston
C.A. Johnstone
C. Kakel
P.H. Kanaar
J.N. Kennedy
H.L.C. King
P.H. Kitson
T.N. Ladefoged
L.F. Lambillion
P.H. Lamsdale
P.N. Lancaster
T.C. Larkworthy
N.W.G. Lattimore
D.M. Lawry
H.J. Laws
W.T. Lawson
C.M. Lavingston
C.E. Lethby
E.T.G. Lewis
R.G. Liddle
N.J. Lock
R. Louisson
W.F. Lovell
A.J. Lusty
E.E. Lyus
R.G. Madders
P.W. Maitland
F.R. Mallon
A.F.X. Mardel-Ferreira
R. Manns
E.G. Marsh
J. Marshall
J.K. Mason
L.D. Matthews
W.G. McDonald
R.L. Meatyard
J.A.H. Medhurst
J.E. Mees
W.W. Merrill
E.B. Merrony
J.A. Messenger
S. Metcalf
C.B. Millbank
J.B. Miller
J.A. Morgan
R.D. Morris
T.F.J. Moss
A.J. Mourilyan
A.E. Musto
D.E. Nicholl
C.F. Norman

F.L. Oldland
A.G. Olley
I.R.W. Olley
G. Osborne
W.J. Pangbourne
C.W. Parker
E.H. Parkes
H.E. Parsons
E.D. Paul
H.J. Paveley
A.R.G. Pearce
C.F. Petch
D.A. Phillips
G.S.P. Phillips
F. Pitkin
B.C. Pocock
L.A. Powell
D. Prangnell
N.G. Pring
M.J.E. Puttock
T.F. Rainsbury
A.H. Randell
R.B.C. Reed
K. Reeve
P.A. Reeve
J.A. Rascorla
R.M. Rickards
K.E.O. Ringdahl
R.W. Robb
H.C. Roberts
H.A. Rockhill
J.H. Rogers
D.R.V. Roper
S.W. Root
A.S. Rush
E.P.S. Russell, A.M.
R.J. Sampson
A.W. Sanders
F.B.F. Sanderson
W.A.R. Sanft
B.S. Sargent
T.R. Saunders
W.H. Seabrook
V.A.L. Shallis
A.C.K. Smith
D.J. Smith
E.A. Smith
G.H. Smith
J.P. Smith
L.A. Smith
L.J. Smith

S.R. Sothcott
A.C. Southwell
C.P. Spedding
J.E. Standen
J.V. Stanley
C.E. Stannard
L.H. Stone
W.H. Stratton
J.H. Sturt
G. Sullivan
E.R. Swift
G.A. Tanner
A.T. Taylor
E.O.M. Taylor
J.T. Taylor
L.J. Thomas
E.O. Thomson
C.S. Tidbury
F.L. Tomlinson
G.A. Tompkins
R.R. Towers
L.A. Tulk
F.G. Turnbull
B.O. Upchurch
N.A. Vernon
F.C. Voaden
J. Waters
R.N. Weir
R.C. West
F.H.J. Weston
T. Wharram
R. Wheeler
A. Whiffen
D.W. White
L.B. Whittam
H.L. Wiggerham
R. Wilcox
J.F. Wilford
E.A. Williams
E.J. Williams
L.A.M. Williams
A.F. Wilson
R.A. Wilson
W.H. Wilson
D.A. Wise, D.S.C.
P.J. Woolf
J.N. Woollard
A.L.O. Wray
R.M. Wright

LONDON FLOTILLA R.N.V.(S)R.

N.R. Armitage
E.H. Beart
Sir M.C. Blennerhassett
R.W. Bristow
W.B. Chisholm
E.A. Collins
J.S.G. Comfort
J.W. Congreve

J.H. Cooper
C.G. Creswick
G.A.R. Darlow
E.A. Dicks
R.A. Durham
C.M. Durno
R.A. Edmunds
A.R. Everitt

M. Eynon
J.M. Foley
J.R. Fradgley
H.A. Gellett
R.H. Gill
T.H. Goodchild
I.B.H. Henderson
R.A. Hill

G.E. Johnson
D. Kemsley
J.L.W. Lane
R.P. Lawrence
W.G. Lewis
R.H. Mead
R.S. Meirick
J.S. Neate

W.A. Pain
T.M. Rumbold
A.V. Stubbs
C.F.B. Stubbs
F.P. Tennyson
B.W.R. Thomas
J.P. Thorpe
G.W. Vaughan

H.R. Walker
B.M.T. Wallace
G.R. Weller, D.S.C.
A.E. White
A.S. Wilson
D.C. Wilson
F.H. Wright
J.O.V. Yates

Lost in H.M.S. Fittleton

20th September, 1976

Sub Lieutenant Christopher Grenfell, R.N.R. – London Division R.N.R.
Chief Radio Supervisor Philip Barber – London Division R.N.R.
Cook Kevin Donoghue – HMS *Superb*
Marine Engineering Mechanic Ian Hewison – London Division R.N.R.
Marine Engineering Mechanic Gerard Hoey – HMS *Sultan*
Radio Operator Richard Massey – Forth Division R.N.R.
Ordnance Electrical Mechanic Charles Newell – Sussex Division R.N.R.
Cook Michael Pilch – London Division R.N.R.
Chief Petty Officer Cook Frederick Pilgrim – London Division R.N.R.
Radio Operator Patrick Quantrill – Southend Communications Training Centre R.N.R.
Communications Yeoman David Skinner – London Division R.N.R.
Mechanician First Class Stanley Turner, BEM – London Division R.N.R.

Notes

Chapter One

1. M. Lewis, *Navy of Britain* (Allen & Unwin, 1948), p. 308.
2. Sir W. Laird Clowes, *The Royal Navy*, Vol. I (S. Low & Marston, 1897), p. 592, and *Dict. Nat. Biog.*, Vol. XX, p. 981.
3. F. C. Bowen, *A History of the Royal Naval Reserve* (Corporation of Lloyds, 1926), p. 2.
4. *ibid.*, p. 3.
5. J. Lennox Kerr and W. Granville, *The R.N.V.R.* (Harrap, 1957), pp. 50–1.

Chapter Two

1. *The Times*, 11 Aug. 1903.
2. *Leader*, 5 Sept. 1907.
3. *Daily Express*, 19 June 1908.
4. R. Hough, *Admiral of the Fleet* (Macmillan, 1969), p.253[alt. Hough, *First Sea Lord* (Allen & Unwin, 1969).]
5. Elveden MSS., 13.

Chapter Three

1. D. Jerrold, *The Royal Naval Division* (Hutchinson, 1927), App. A., p. 333.
2. Sir John Smyth, *Great Stories of the Victoria Cross* (Barker, 1977), pp. 90–1.
3. Lieut. Stanley Geary, *The Collingwood Battalion, Royal Naval Division*, (private, 1916), *passim*. A large profusely-illustrated monograph of 44 pp., printed in Hastings, containing biographical notes of all the officers and a list of the men.
4. D. Jerrold, *The Hawke Battalion* (Ernest Benn, 1925), has a fine photograph of Lieut. C. S. Codner, M.C., and his beard, facing p. 182.
5. *Memoir and Poems of A. W. St. C. Tisdall, V.C.* (Sidgwick & Jackson, 1916).

Chapter Four

1. *Morning Post*, 30 May 1921.
2. *Daily Express*, 25 May 1926.
3. *Sunday Express*, 14 April 1935.
4. *Manchester Guardian*, 31 August 1937.

Chapter Five

1. *See* G. G. Connell, *Valiant Quartet* (W. Kimber, 1979), pp. 28–32 & p. 40.
2. *ibid.*, pp. 89–90.
3. A. Raven and J. Roberts, *British Cruisers of World War Two* (Arms & Armour Press, 1980), p. 335 (caption).
4. S. W. Roskill, *The War at Sea 1939–45*, Vol. II (H.M.S.O., 1956), p. 212.
5. Kerr and Granville, *op. cit.*, pp. 178–9.
6. The author.

Chapter Six

1. The author.
2. The Senior Chaplain present, the Rev. G. C. Cutcher, R.N.R., had been baptised by Tisdall's father, the Rev. W. St. Clair Tisdall, at St. George's, Deal.
3. The R.N. Division Association held its annual lunch on board the *Chrysanthemum* until 1980, when advancing age supervened. The name-plate of the locomotive named *The Royal Naval Division* has an honoured place in the *President*.

Index